OXFORD WORLD'S CLASSICS

THE NATURAL HISTORY OF SELBORNE

GILBERT WHITE was born in Selborne in the county of Hampshire in 1720. After a country childhood and study with a tutor in Basingstoke he went to Oriel College, Oxford, where his reading and study of the classics was to have an impact on his later development as a writer. He was ordained in 1749, and although he never became vicar of Selborne, he held the curacy there from 1784 until his death in 1793. In 1751 he began a 'Garden Kalendar', and started the improvements of the grounds of the White family home, The Wakes, that were to continue for forty years. He completed an annual study of the flora of the parish in 1766, and from 1768 he kept journals of his natural history observations. His correspondence with the eminent naturalists Thomas Pennant and Daines Barrington eventually resulted in *The Natural History of Selborne* (1789), his only published book. He died a few years later, in June 1793. His home at Selborne now houses a museum in his memory.

ANNE SECORD is an Editor of *The Correspondence of Charles Darwin*, and an Affiliated Research Scholar in the Department of History and Philosophy of Science, University of Cambridge.

OXFORD WORLD'S CLASSICS

*For over 100 years Oxford World's Classics have brought
readers closer to the world's great literature. Now with over 700
titles—from the 4,000-year-old myths of Mesopotamia to the
twentieth century's greatest novels—the series makes available
lesser-known as well as celebrated writing.*

*The pocket-sized hardbacks of the early years contained
introductions by Virginia Woolf, T. S. Eliot, Graham Greene,
and other literary figures which enriched the experience of reading.
Today the series is recognized for its fine scholarship and
reliability in texts that span world literature, drama and poetry,
religion, philosophy, and politics. Each edition includes perceptive
commentary and essential background information to meet the
changing needs of readers.*

OXFORD WORLD'S CLASSICS

GILBERT WHITE

The Natural History of Selborne

Edited with an Introduction and Notes by
ANNE SECORD

OXFORD
UNIVERSITY PRESS

OXFORD

UNIVERSITY PRESS

Great Clarendon Street, Oxford, OX2 6DP,
United Kingdom

Oxford University Press is a department of the University of Oxford.
It furthers the University's objective of excellence in research, scholarship,
and education by publishing worldwide. Oxford is a registered trade mark of
Oxford University Press in the UK and in certain other countries

First published 2013
First published as an Oxford World's Classics paperback 2016

Impression: 5

Published in the United States of America by Oxford University Press
198 Madison Avenue, New York, NY 10016, United States of America

British Library Cataloguing in Publication Data

Data available

Library of Congress Control Number: 2015941606

ISBN 978–0–19–873775–9

Printed in Great Britain by
Clays Ltd, Elcograf S.p.A.

CONTENTS

LIST OF ILLUSTRATIONS

Unless otherwise indicated, all illustrations are taken from Thomas Pennant's *British Zoology*, 2nd edn., iv (1770), courtesy of the Bodleian Libraries, the University of Oxford (8° P. 16. Linc).

INTRODUCTION

The Natural History of Selborne is so established a classic that it is hard to take seriously its author's fears that as tastes changed the book might quickly sink into oblivion. Instead, the work has never lost its ability to attract new readers. Over the past two and a quarter centuries, editors and readers have been responsible not only for keeping the work in print but also for establishing the reputation of its author, the Reverend Gilbert White, as the producer of a new kind of natural history. The significance of White's book has been recast many times and its originality redefined by later scientific questions, but, paradoxically, the very elements that characterize it as a thoroughly eighteenth-century work have proved to be the essence of its enduring, widespread appeal. *The Natural History of Selborne* has continued to speak to readers because what it has to say is only a part of its attraction. It is in the manner of its telling that much of the charm of the work lies.

White's decision to depict the natural history of his own parish in a series of letters addressed to the naturalists Thomas Pennant and Daines Barrington has proved essential to its longevity. Employing the sense of immediacy that letters allow, White not only sparks our interest with those 'anecdotes' of animal behaviour he thought so essential to a work of natural history, but also teaches us one of the subtlest of lessons: how to observe. His accounts of natural phenomena and the processes of nature are inseparable from the observational practices most appropriate for their elucidation. No sooner are we invited to gaze upon the 'heavy and bloated' viper basking in the sun than the reptile has been cut open to reveal the live young in her abdomen (p. 168). A naturalist, he tells us, must develop the ability to identify a bird in flight from the most fleeting of glances, while the utmost patience is required for observing the excavation of a place to hibernate by a tortoise, whose legs move almost as slowly as 'the hour-hand of a clock' (p. 123).

White's two series of letters, with their combination of detached analysis and delight in the ordinary occurrences of nature, were carefully compiled to pique and maintain his readers' interest, in the hope that this would encourage them to attend to the less gripping issues of taxonomy and systematic record keeping.[1] Such skills were a necessary attainment if White's wish that his book might stimulate others to the scientific observation of their own localities was to be fulfilled. By focusing attention on Selborne alone, White was not limiting the reach of his work but expanding it. His *Natural History*, he hoped, would serve as a model for how best to gain and usefully apply a knowledge of nature. 'Men that undertake only one district', he pointed out, 'are much more likely to advance natural knowledge than those that grasp at more than they can possibly be acquainted with' (p. 110). In White's view, it was these 'partial writers', rather than 'more general' ones, who 'by degrees may pave the way' to nothing less than 'an universal correct natural history' (p. 71).

A Parochial History

It is in the light of White's grand vision of a 'universal correct natural history' that we can understand his concern that the public would laugh at an old country parson's book. During the years that he corresponded with Pennant and Barrington, and while he prepared these letters for publication, White remained unshaken in his conviction that a 'correct' natural history depended upon showing not only that the application of natural knowledge could benefit society but also that social hierarchies, and the authority assumed by the clergy and the gentry, were part of the natural order of things. Gentlemen were qualified to pursue rational inquiries into nature by virtue of their education and their leisure—a role that White, an Oxford-educated clergyman from a family deeply embedded in the Anglican church, played out in Selborne, a parish of 12½ square miles with 700 inhabitants, in which he lived almost his entire life. Months after *Selborne* was published, however, the French Revolution and the social unrest it produced in England were the first serious assaults on the authority assumed by the clergy and the upper orders of society. White feared that this political upheaval would disrupt the natural order of things—the belief that divine governance, discernible through the investigation of nature, showed how the correct functioning of the whole depended upon the adaptation of the parts. The destruction of such authority would indeed make a book written

¹ Lucy B. Maddox, 'Gilbert White and the Politics of Natural History', *Eighteenth-Century Life*, 10 (1986), 48.

by an old country parson laughable. In the event, the troubled times after 1789 probably made White's intended audience of the clergy and the gentry more receptive to his work. 'You cannot abhor the dangerous doctrines of levellers and republicans more than I do!', he wrote to an admirer of his book. 'I was born and bred a Gentleman, and hope I shall be allowed to die such.' He consoled himself with the thought that the 'common people' of Selborne were 'simple-minded and know nothing of Jacobin clubs', and he continued to draw comfort from recording the perceptible order and regularity of natural phenomena until a few days before his death in 1793.[2]

White had no need to be as overt as this in *Selborne*, which was completed before the French Revolution and, as a lavish and expensive work, was addressed an audience he could assume would share his opinion that they alone were qualified to join together to produce a universal natural history. His patrician views of nature, however, are clear from the text itself. *Selborne* is shaped by his love of the classical authors, of John Milton's poetry, and of physico-theological works. It was the entanglement of these literary and religious resources with White's acute observation of nature that enabled him to convey feelings of wonder or awe as well as provide a philosophical description of natural phenomena. *Selborne* is also an expression of cosmopolitanism, and of White's participation in a scientific culture with global interests, exhibited not least in the number of travel narratives mentioned in the text and by his references to far-flung places, from China to Peru.[3]

White's own interest in the natural world is itself a parochial history. His early life revolved around Selborne and Oxford, but when his ambition for an academic life came to nought in 1757 with his failure to be elected Provost of Oriel College, and after he inherited The Wakes (the White family home in Selborne) in 1763, he was reluctant to move away from the small village he had known since he was a child. He had already devoted years to making fashionable improvements to the gardens at The Wakes, recording his activities and their results in the 'Garden Kalendar', which he started in 1751. But it was his duties as a clergyman, combined with his decision to

[2] Rashleigh Holt-White, *The Life and Letters of Gilbert White of Selborne*, 2 vols. (London, 1901), ii, 255.

[3] Tobias Menely, 'Traveling in Place: Gilbert White's Cosmopolitan Parochialism', *Eighteenth-Century Life*, 28 (2004), 46–65.

remain in Selborne, that developed and intensified his interest in natural history.

At a time when most people were born and died in the same village, an eighteenth-century parish was not only a geographical area under ecclesiastical administration but was also—in part through the role of the Church—constituted by its social allegiances, the duties between rich and poor, and the relationship of the inhabitants with the lands and creatures within its bounds. White was concerned to introduce improvements that would benefit the occupants of the parish of Selborne by showing how the rural economy (the sowing of crops and the raising of animals) could be guided by a natural calendar that revealed the connections between the various phenomena of nature. He had been inspired by the Reverend Benjamin Stillingfleet, who had articulated the same aim in his 'Calendar of Flora' of Norfolk, which he published in 1762 with a translation of a calendar of the flora of Uppsala by Alexander Malachias Berger, one of the followers of the Swedish naturalist Carl Linnaeus, as 'Calendar of Flora, Swedish and English'.[4] Stillingfleet's conception of the links between an understanding of nature and the improvement to people's daily lives not only prompted White to start his own calendar of the flora of Selborne in 1766, but also allowed him to see how his religious vocation and life as a clergyman could come together with his love of natural history.[5] Viewing his pursuit of natural history in this light might also have provided White with a justification for choosing to remain a curate in Hampshire instead of advancing his clerical career by becoming a vicar elsewhere.

A full understanding of natural events that consistently accompany one another, such as the correlation between the appearance of certain birds and the germination of particular plants, could be achieved only by making observations in the same place over long periods of time and by comparing the observations of naturalists in different places. White's attentiveness to what was happening around him and the possible connections between natural events that might be detected by noting 'co-incidences', was honed by his practice of travelling everywhere on horseback, and particularly by the

---

[4] Benjamin Stillingfleet, *Miscellaneous Tracts in Natural History*, 2nd edn. (London, 1762), 229–316.

[5] Paul Foster, *Gilbert White and his Records: A Scientific Biography* (London, 1988), 62–8.

opportunity for repeated observations afforded by his journeys along the same routes when going to preach sermons in nearby churches. Natural history observations had already crept into his 'Garden Kalendar', and by the time White began his 'Flora Selborniensis' in 1766 he was well primed to follow Stillingfleet in not confining it narrowly to plants but including 'some co-incidences of the coming, & departures of birds of passage, & insects; & the appearing of Reptiles'.[6]

These records contributed to *The Natural History of Selborne*, and were essential to White's development as a naturalist, but his ability to produce such a work owed far more to the transformative events of 1767 that marked his entry into the scientific community. In this year, he met, on one of his many visits to London, the scientific traveller and soon-to-be circumnavigator of the globe, Joseph Banks; he began a correspondence with the zoologist Thomas Pennant; and he received, as a gift, a new kind of notebook for recording observations: *The Naturalist's Journal.*

The contact between Pennant and White had been established through Pennant's publisher—White's brother Benjamin White, the leading natural history bookseller in London. Pennant, a wealthy Welsh antiquary, zoologist, traveller, and newly elected Fellow of the Royal Society of London, was preparing a second edition of his *British Zoology*, and realized that White could provide reliable information. The gain, however, was not all on his side. Both White and Pennant appreciated having a correspondent in a distant part of the country, and in exchange for White's information, the well-connected Pennant could provide contacts with a wider scientific community. Recognizing the importance of local natural histories for the general works he was publishing, it was probably Pennant who encouraged the barrister, antiquarian, and naturalist Daines Barrington to send White a copy of his newly published *Naturalist's Journal*.

Barrington, who, like White, was inspired by Stillingfleet's suggestion that naturalists should note the comings and goings of birds and the cyclical occurrence of other natural phenomena, had understood that a method of organization was required to encourage the making of such records. To this end, he designed *The Naturalist's*

[6] 'Flora Selborniensis', *The Journals of Gilbert White, 1751–93*, ed. Francesca Greenoak, 3 vols. (London, 1986–9), i. 188.

Journal, which consisted of a set of printed forms, each page covering a week, for the systematic recording of a variety of natural phenomena over a year. 'Many . . . particulars will daily offer themselves to the observer', Barrington stated in the preface, and 'when his attention to such points hath once become habitual, and from many such journals kept in different parts of the kingdom, perhaps the very best and accurate materials for a General Natural History of Great Britain may in time be expected'.[7]

White immediately took to *The Naturalist's Journal*, and although his entries soon began to spill over the ruled columns (and by the 1770s he was adding blank pages for more extended notes), this was the form in which he recorded his observations for the rest of his life. Although designed to be a record of the natural world, White's journal also bore witness to the times he lived through. There are events of such enormity that White seemingly could not refrain from letting them burst in. It is in the journal that he noted the outbreak of wars, and, on 21 January 1793, between the turnip greens coming in and the snowdrops in bloom, he recorded the beheading of Louis XVI by French revolutionaries, adding, in evident shock, that the king's body was 'flung into a deep grave without any coffin, or funeral service performed'.[8]

The journal also records the comings and goings of White's visitors, and the births, deaths, and marriages in his large extended family. It is an appropriate repository for these records, because beyond Pennant and Barrington, White's main sources of natural history observations were his relatives. In particular his brother John, army chaplain in Gibraltar, was able to provide information and observations to compare with those of Selborne, especially on the appearance and disappearance of birds. John White, along with other relatives and friends, is not named in *Selborne*, appearing only as an 'accurate observer' or, occasionally, 'my brother' in Andalusia. As far as scientific discussions went, it was Pennant and Barrington who were centre stage. To White's regret, neither man ever visited Selborne. Although they met occasionally in London, White's relationships with them were maintained largely through correspondence.

---

[7] [Daines Barrington], *The Naturalist's Journal* (London, 1767), preface.
[8] Greenoak (ed.), *The Journals of Gilbert White, 1751–93*, iii, 446.

Tradition and Novelty

White's decision to publish a natural history made up of his correspondence built on the success of the first of his monographs on martins, swallows, and swifts, written as a letter to Barrington at his request and presented at the Royal Society of London in February 1774. Soon afterwards White outlined his plan for a book: material from his journals and letters would produce a 'moderate volume', to which he would add 'some circumstances of the country' and 'its few antiquities', that is, its monuments and ancient records.[9] Despite the fears of White's closest friend, the Reverend John Mulso, that the natural history was at risk of being 'overwhelmed by the Rubbish of the Antiquities',[10] in the fourteen years it took before the work was completed and sent to Benjamin White's publishing house, White's plan did not change. When the book appeared in November 1788, it had two parts and bore the title *The Natural History and Antiquities of Selborne*, echoing the well-known genre of county histories such as John Aubrey's *The Natural History and Antiquities of the County of Surrey* (1719) and John Wallis's *The Natural History and Antiquities of Northumberland* (1769). In choosing an area as small as a parish, however, White selected a unit of study that from the publication of White Kennett's *Parochial Antiquities* (1695) had been confined to antiquarian pursuits. For that reason alone, White perhaps felt obliged to insist in the 'Advertisement' to his book that a parochial history 'ought to consist of natural productions and occurrences as well as antiquities'.

White, however, was not alone in calling for this kind of history. In 1780 the first number of the *Bibliotheca topographica Britannica* included 'Queries for the better Illustrating the Antiquities and Natural History of Great Britain and Ireland'. Drawing on a tradition of parochial questions drawn up by antiquaries, from which knowledge of larger areas could be built up, the 'Queries' encouraged individuals to investigate their parishes thoroughly, but warned that it took 'long practice and observation' to answer the natural history questions. These included accurately describing the soils, topography, scenery, history, and statistics of a parish—topics that were dealt with in the first nine letters of *The Natural History of Selborne*,

[9] Holt-White, *Life and Letters of Gilbert White*, i, 250.
[10] Holt-White (ed.), *Letters to Gilbert White . . . from . . . John Mulso*, 285.

which White composed especially for the book. Many of the subjects covered in White's *Natural History* are also listed as worthy of notice in the 1780 'Queries', including 'remarkable echoes' and superstitious beliefs, and the suggestion is made that readers should record observations on the plan of Barrington's *Naturalist's Journal*.

If, by the time it was completed, White's book appeared to follow an established tradition, what made an early reviewer hail *Selborne* as one of the most original works ever published? From the start, readers were especially attracted by the natural history, which they admired not only for the information it contained but also for 'an Originality in the Management of it that is very pleasing'.[11] This management consisted of conveying the novelty of 'out-door' observations of the behaviour of animals—what White considered the 'life and soul' of natural history—through the medium of letters.

White's use of the epistolary form was not new. Communications to scientific societies and publication in periodicals usually took the form of correspondence. Moreover, private exchanges between individual naturalists had appeared in print, such as William Derham's edition of *Philosophical Letters between the Late Learned Mr. Ray and Several of his Ingenious Correspondents* (1718), with which White was familiar. From the mid-eighteenth century, following the popularity of Samuel Richardson's epistolary novels *Pamela* (1740–1) and *Clarissa* (1748–9), letters as a literary device were also associated with heightened emotion. Dramatic effect was enhanced by characters writing about events as they were happening, thereby increasing the reader's sense of inhabiting the present moment. These novels depended not on influencing the passions of readers but on developing 'sentiment', a moral and aesthetic category that stimulated the pleasures of the imagination. Although the appreciation of scientific work was usually achieved through reason not sentiment, the originality of White's *Selborne* consisted in its ability to spark both faculties in its readers: 'not only the understanding is informed,' a reviewer pointed out, 'but the imagination touched'.[12]

A similar assessment of White's letters had already been made, long before *Selborne*, by the young physician, John Aikin, who held that with respect to 'poetical effect' there was little to trump the

[11] Holt-White (ed.), *Letters to Gilbert White . . . from . . . John Mulso*, 346.
[12] *The Topographer*, I (1789), 40 (see Appendix II, p. 242).

'grandeur' which 'cool science and demonstration offer to the enlight-ened mind'.[13] According to Aikin, natural history sources such as the extracts from White's correspondence on migration that had been published in the second edition of Pennant's *British Zoology* in 1768, provided a quality more important than genius in poetry—namely, novelty.[14] Novelty stimulated the imagination, and the pleasures associated with it included agreeable surprise, the attainment of knowledge, the gratification of curiosity, and the experience of vari-ety. Mulso, though no naturalist himself, had also recognized this quality in natural history, telling White that he was 'not at all sur-prized at the Pleasure you take in your Pursuits of natural knowledge; very far from it; I know nothing more capable of satisfying the Curiosity of the human Mind, which is allways searching after Novelty, for the Subject is inexhaustible'.[15] And White informed his readers in turn that natural history can be an inexhaustible source of novelty for anyone, because 'all nature is so full, that that district produces the greatest variety which is the most examined' (p. 45).

The novelty of White's natural history lies in the clarity with which he showed the importance of appreciating not only the kind of attention an object invites but also the manner most appropriate for expressing what is observed. The systematic language for ordering plants and animals developed by Linnaeus and others, and published between 1753 and 1758, greatly facilitated the categorization and classification of natural objects beyond local boundaries, and White agreed that 'there is nothing to be done in the wide boundless field of natural history without system'.[16] But description of this sort was not his aim. White wanted instead to find a method by which the 'life and conversations' of animals could be expressed. The 'stile' of a work, he believed, should be adapted to the subject in all cases.[17] His practice reflects the essayist Joseph Addison's observation that 'not only what is Great, Strange or Beautiful, but any Thing that is Disagreeable when look'd upon, pleases us in an apt Description'.[18] White, with

[13] John Aikin, 'An Essay on the Plan and Character of the Poem', in James Thomson, *The Seasons* (London, 1778), xv.

[14] John Aikin, *An Essay on the Application of Natural Philosophy to Poetry* (Warrington, 1777), 4, 124–5.

[15] Holt-White (ed.), *Letters to Gilbert White . . . from . . . John Mulso*, 215–16.

[16] Holt-White (ed.), *Life and Letters of Gilbert White*, i, 179.

[17] Holt-White (ed.), *Life and Letters of Gilbert White*, i, 192

[18] Joseph Addison, ['The Pleasures of the Imagination'], *Spectator* (30 June 1712), 1.

great skill, translated the practices of both the observer and the observed into striking descriptions, in order to bring the natural history of the parish to life under the very eyes of his readers. Unlike the nearly exclusive emphasis placed on vision in taxonomic descriptions, White's accounts make clear that other senses are also necessary when observing nature. His natural history comes alive through vivid accounts of rancid-smelling bats, stinking snakes, the putrid stench of death, and the sulphurous smell of a blue mist that heralds a thunderstorm. We learn not only that some species of birds must be distinguished by their song but also that certain topographical features produce remarkable echoes (a feature White investigated in Selborne by bawling out Latin phrases). He vividly calls to mind the effects of weather by describing the excessive heat of the summer of 1783 and the biting cold of a winter so severe that the contents of chamber pots froze under beds. The pleasures of the imagination are heightened, as Addison had argued, when the ideas of two senses 'recommend each other'.[19]

Observing Wonder

White observed with an outdoor naturalist's eye wherever he was and regardless of whether what he was viewing was domestic, exotic, or wild. In *Selborne*, he reports that his parlour cat was excessively electric on two 'Siberian days' in a harsh winter, and that his aunt's tortoise had a dislike of rain. His interest in watching the actions of fish in a glass bowl while visiting his brother Henry was crowned by his witnessing that phenomenon all too familiar to the owners of goldfish: 'It was here', he tells us, 'that I first observed the manner in which fishes die' (p. 211). His memorable visit to see the Duke of Richmond's moose was made in the hope—disappointed, in the event—of seeing a live animal, and on reading Pennant's account of herons' nests at Cressi Hall he exclaimed: 'Fourscore nests of such a bird on one tree is a rarity which I would ride half as many miles to have a sight of' (p. 49). When viewing 'astonishing collections' of live birds from places as distant as South America or the Guinea coast, he noticed that they all had hard bills. This, he deduced, was because it was impossible to transport soft-billed birds, whose diet of

---

[19] Joseph Addison, ['The Pleasures of the Imagination'], *Spectator* (23 June 1712), 2.

worms, insects or berries could not be provided on board ships (unlike the seeds eaten by hard-billed birds). Appreciating that he could not always see live animals, White visited London to see 'strange and wonderful specimens', and he eagerly anticipated the discoveries that the naturalist Joseph Banks would make during his voyage around the world.

Given his experience of such wondrous sights, it might be thought that White would describe the everyday occurrences of nature in Selborne in a different way. But his perspective does not banish wonder, it relocates it in the ordinary. The appearance of what he took to be the same species of crocus in both the spring and the autumn is, in his view, 'one of the wonders of the creation, little noticed, because a common occurrence; yet ought not to be overlooked on account of its being familiar, since it would be as difficult to be explained as the most stupendous phaenomenon in nature' (p. 187). His observations on the lowly earthworm overturned its reputation as a garden pest (though not the belief that it was 'addicted to venery'; p. 173) by demonstrating its essential role in the natural economy. White's attention to the processes of life led him to write about everything as a curiosity, and sometimes as a mystery that exceeded the limits of observation. He provided a descriptive layer of observation of the honey buzzard's egg, for example, but opening the egg to view the embryo halted the very process of life upon which an understanding of the much debated topic of generation (the coming into existence of new individuals) depended. As for eels, whose generation 'is very dark and mysterious', White was not able even to determine whether the 'threads' sometimes discovered in them were their young. Equally dark was the method by which toads 'procreate and bring forth'.

Other methods of procreation could be observed, even if they remained obscure. 'The copulation of frogs (or at least the appearance of it; for Swammerdam proves that the male has no *penis intrans*) is notorious to every body', White pointed out, 'because we see them sticking upon each other's backs for a month together in the spring' (p. 40). As easily observable was the ponderous tortoise, whose amorous feelings prompted relatively energetic rambles on tiptoe but who was 'said to be a whole month in performing one feat of copulation' (p. 123). In contrast, no one before White had had sufficiently skilled eyes to detect that swifts 'tread, or copulate, on the wing', the loud, piercing shriek a pair utters when sinking through the air indicating, in White's view, the 'juncture when the business of generation is carrying on' (p. 147). White's ability to reason from what he could observe on this question derived from his long experience in watching birds' behaviour.

White's investigation of mysterious phenomena, his records of everyday occurrences, and his interest in unusual events and exotic animals, do not appear in *The Natural History of Selborne* as distinct categories, but as parts of an interconnected whole, illustrative of living nature as a scene of constant change and process. White's text emphasizes not only the wonder produced by the detection of coincidences and connections in nature but also the methods an observer must learn in order to notice them. The investigation of the 'life and conversation' of animals, White warned, is 'not to be attained but by the active and inquisitive, and by those that reside much in the country' (p. 120). It required an attentive eye to disentangle particular phenomena from the mass of seemingly unconnected events that present themselves to an observer of nature. The apparently random organization of the subject matter of *Selborne* does, however, reflect a temporal order that was central to White's aims of demonstrating what to attend to when observing natural processes. The chronological arrangement of the letters, from 1767 to 1787, shows the slow accumulation of observations through the passing of the seasons, changes in the landscape, the new questions that arise during the course of investigations, and the development of White's relations with other observers.

In order to focus the attention of his readers, White carefully manipulated his materials when constructing *Selborne*. He concealed the fact that not all the letters in the book were genuine letters that had been sent to Pennant and Barrington. Some had been addressed to other correspondents, others were based on the translation of entries from White's journals into letter form, and even those that had originally been sent to the two naturalists were divided up into several shorter ones. The result is that most of the letters address only a single topic. This focus was not unusual among eighteenth-century naturalists, who believed that attention could be concentrated sufficiently for exact observation only if no more than one object at a time was considered; moreover, the difficulty of sustaining such intense attention required observers to return to the same object repeatedly, selecting different details or aspects each time, and confirming what had already been observed. In addition, the reader's experience of moving from one subject to another in successive letters was conducive to the pleasure of novelty and the refreshment provided by variety, by which means, according to Addison's aesthetic theory, 'the Mind is every Instant called off to something new, and the Attention not suffered to dwell too long, and waste itself on any particular Object'.[20] Moreover, by jumping from one topic to another in the same way as phenomena are encountered in nature, the published letters echo the unpredictability of observation, where even anticipated phenomena, if detectable only by the practised eye, could be easily missed on any one day, and others could occur unexpectedly. The epistolary mode, with its ability to convey the immediacy of experience, possesses the potential not only to imitate the way nature appears to an outdoor naturalist but also to underline the need to take advantage of all occasions for observation.

A Community of Observers

When first encouraged to write on natural history, White had pointed out that it was

no small undertaking for a man unsupported and alone to begin a natural history from his own autopsia! Though there is endless room for observation in the field of nature, which is boundless, yet investigation (where a man endeavours to be sure of his facts) can make but slow progress; and all that one could collect in many years would go into a very narrow compass. (pp. 106–7)

By deciding to publish a compilation of his letters, White turned what he thought was a disadvantage—his reliance upon his own observations—into an essential feature of his *Natural History*. Although he bemoaned the fact that he had no like-minded companion in his neighbourhood to 'quicken my industry and sharpen my attention', far from presenting himself as a solitary observer of nature he positioned himself as one link in a community of investigators. His decision to present separately the two correspondences that comprise the book (rather than interleaving them chronologically) points to a continuous process of collaboration in natural history, in which

[20] Addison, *Spectator* (23 June 1712), 1.

'candour and openness' result in knowledge that transcends individual experience.

Strong though White's persona is in *Selborne*, it is possible to detect shifts in his voice depending on which correspondent he is addressing—a variation in tone that would have been more apparent to contemporary readers, to whom Pennant and Barrington were well-known figures, than it is today. This is most obvious in his discussions of Pennant's arguments for, and Barrington's against, bird migration. White's own investigation of the subject was based on his meticulous records of the annual appearances and disappearances of birds. So 'mysterious and wonderful' was the early departure of swifts in the 'sweetest season in the year', long before they might be affected by heat or lack of food, that it 'not only baffles our searches,' White exclaimed, 'but almost eludes our guesses!' (p. 150). The evidence White gathered on the subject suggested that in some cases the arguments for migration were as difficult to accept as the notion that birds hid themselves away in a torpid state during the winter. Although White believed that many birds did migrate, when he returned to the strange case of swifts he used observable changes in their appearance to present a case against migration. When the swift first arrives in the spring, White noted that it is

a dark sooty colour; yet by being for 16 hours together almost constantly in the sun and air, it becomes before its departure much bleached, and as it were what the country people call piss-burnt, like an old weather-beaten brown wig: and yet it returns sooty again in the Spring. Now if they go into warm regions during our winter, why do they not return sun-burnt, as they went off? It is a matter of curious enquiry to consider when swifts moult. Change their feathers here they certainly do not: and if they have as much occasion for their wings while absent, as while here, they would find no opportunity to spare several feathers at a time.[21]

Such considerations led White to 'suspect that they hide, rather than migrate' over the winter, although he and other British naturalists resisted Linnaeus's view that members of the swallow family hibernate at the bottoms of lakes. White's reasoning shows how he weighed up and combined visual evidence and conjecture on a phenomenon that could not be witnessed directly.

The passage above was originally part of letter 19 to Pennant, but

---

[21] Holt-White, *Life and Letters of Gilbert White*, i, 164.

it was omitted when the manuscript was cut down to just one topic. This was probably because White had included an almost identical passage in letter 21 to Barrington, which was all about swifts. Removing the potential evidence for the hibernation of swifts from the letter to Pennant and leaving it in one addressed to Barrington, who was 'no great friend to migration' (p. 116), also reduced any suggestion of disagreement without undermining the essential task of observers to take account of all views on unsolved questions. Letters, which could be read as private communications, even when made public, lent themselves to the discussion of provisional knowledge, and served to encourage others to look, check, and, if necessary, correct—as many future editors of *Selborne* were to do well into the nineteenth century. By presenting different views in his *Natural History*, White showed the necessity of not coming to premature or hasty conclusions. Laying out all sides of an undecided issue also allowed readers not only to judge for themselves but also to see the methods by which to gather evidence. White never resolved the question of whether some birds hibernate, and he continued to search for evidence that house martins and swallows did partake of this 'benumbed state' in 'secret dormitories' in Selborne. His last attempt to discover torpid swallows was made just weeks before his death.

The vexed question of migration, which required White to look beyond Selborne, also reveals how he went about recruiting observers and obtaining reliable information. He tells Pennant, who was eager for information on the migration of the stone-curlew, that he knows a gentleman farmer who 'is abroad early and late' and will therefore be 'a very proper spy upon the motions of these birds; and besides, as I have prevailed on him to buy the *Naturalist's Journal* (with which he is much delighted), I shall expect that he will be very exact in his dates' (p. 47). Although this friend was 'no naturalist', two years later the notes he had made in the journal enabled him to provide White with information on these birds.

So familiar has this investigative, communal method become that it is hard to appreciate its novelty at the time. But White's use of it shows that it served a far more fundamental purpose than simple information gathering. Record-keeping based on first-hand observation, shared with others, meant that one could begin to distinguish between superstition and observable phenomena. These were methods familiar to scientific men at least since the origins of the Royal

Society in the 1660s, but they had never been applied so thoroughly or consistently in the close observation of a single district. In some cases, anecdote and superstition were found to be valuable sources of information about the natural world. When White experimented with a tame bat and discovered that it would eat raw meat, he concluded that 'the notion, that bats go down chimnies and gnaw men's bacon, seems no improbable story' (p. 28). He was, however, dismayed by the superstitious panic exhibited by the common people during the summer of 1783, although he did concede that 'the red, louring aspect of the sun' and the intense heat (due to the effects of a volcanic eruption in Iceland) gave 'reason for the most enlightened person to be apprehensive' (p. 233). Such an occurrence, 'unlike any thing known within the memory of man', called into question faith in the providential regularity of the natural order; but White maintained that systematic observation over a long period of time would show the seasonality of even what seemed extreme or abnormal weather. *The Natural History of Selborne* aimed to reveal the harmony and balance of the natural economy, ranging from the benign fertility of the English soil to the hospitable climate.

Providence

White's attention to the interconnectedness of nature has led to an interest in the relevance of his work to ecology, but to associate White with so recent a discipline obscures his providential view of nature and his aim, following the seventeenth-century naturalist John Ray, whom he admired greatly, of revealing the wisdom of God manifested in the works of the creation.[22] The dominant mode of discourse about natural objects in the eighteenth century was in terms of their place in the great chain of being—the idea that all beings and matter are organized in a strict hierarchical structure from the most lowly and least complex minerals up to the highest order of angels, and ultimately God—and the notion of a natural economy, or means by which the balance of nature was maintained through powers placed in the world by God. These concepts provided White with the interpretative framework for his natural history, and governed his

---

[22] John Ray had published *The Wisdom of God Manifested in the Works of the Creation* in 1691.

investigations of those 'two great motives which regulate the proceedings of the brute creation': love and hunger (p. 121). His belief in providence and the caring superintendence of a benevolent God is most apparent in those situations that appeared least beneficial. Fish were able to survive in a sterile pond, for example, because cattle, liking to stand in ponds during the heat of summer days, 'drop much dung, in which insects nestle; and so supply food for the fish, which would be poorly subsisted but from this contingency. Thus Nature, who is a great economist, converts the recreation of one animal to the support of another!' (p. 21). During the remarkable cold of January 1776, White similarly noted with approval that 'crows watched horses as they passed, and greedily devoured what dropped from them' (p. 227).

White is equally observant of the behaviour of birds in the 'season of love', from the 'toying and playful manner' of ring doves to the goatsucker's glancing 'in the dusk over the tops of trees like a meteor' (p. 189). How 'strange and ridiculous', he tells us, is the 'amorous sound of a crow', while rooks, in the breeding season, 'attempt sometimes in the gaiety of their hearts to sing, but with no great success' (p. 191). We also learn of a sagacious sow, who, 'when she found occasion to converse with a boar . . . used to open all the intervening gates, and march, by herself, up to a distant farm where one was kept; and when her purpose was served would return by the same means' (p. 170). It was, however, maternal affection above all that 'sublimes the passions, quickens the invention, and sharpens the sagacity of the brute creation' (p. 124). *Selborne* provides many examples of such behaviour.

But White's views on maternal affection also presented the greatest challenges to a providential view of nature. He considers that the behaviour of the cuckoo in dropping its eggs into other birds' nests 'as it were by chance, is such a monstrous outrage on maternal affection, one of the first great dictates of nature . . . that, had it only been related of a bird in the Brasils, or Peru, it would never have merited our belief'. His faith in nature's providential organization is restored only by Barrington's observation that 'the cuckoo does not deposit its egg indiscriminately in the nest of the first bird that comes in its way, but probably looks out a nurse in some degree congenerous, with whom to intrust its young'. This suggestion, White declares, 'is perfectly new to me; and struck me so forcibly, that I naturally fell into a train of thought that led me to consider whether the fact was so, and what reason there was for it' (p. 104). White's inquiries showed that cuckoos in Selborne appeared only in the nests of insectivorous, soft-billed birds, leading him to doubt Francis Willughby's reported sighting of one in the nest of a hard-billed bird, since it could not tolerate a diet of acorns and grains. But should it appear that the cuckoo, which lacks the natural affection

that seems to raise the kind in general above themselves, and inspire them with extraordinary degrees of cunning and address, may be still endued with a more enlarged faculty of discerning what species are suitable and congenerous nursing-mothers for its disregarded eggs and young, and may deposit them only under *their* care, this would be adding wonder to wonder, and instancing, in a fresh manner, that the methods of Providence are not subjected to any mode or rule, but astonish us in new lights, and in various and changeable appearances. (pp. 104–5)

If investigating the lack of maternal affection in the cuckoo provided White with a forceful argument for the limitless and unpredictable power of providence, he was unable to discern this in another 'monstrous perversion' that led some female pigs, dogs, and cats to devour their young because their owners had handled them too freely or had moved them. This 'horrid and preposterous murder' was less easy to comprehend, in White's view, than the destruction of children by an abandoned mother, since humans are capable of any enormity if reason is perverted and 'bad passions let loose'. But 'why the parental feelings of brutes, that usually flow in one most uniform tenor, should sometimes be so extravagantly diverted' was a problem that White had to 'leave to abler philosophers than myself to determine' (p. 126).

The Legacy of The Natural History of Selborne

White's puzzles did indeed serve to stimulate later readers of his work. The naturalist Edward Blyth, for example, announced in his 1836 edition of *Selborne* that he had arrived at the 'true solution' of the problem of cuckoo behaviour.[23] While experts added their knowledge to that of White in a constant stream of new editions from

[23] Edward Blyth (ed.), *The Natural History of Selborne, with its Antiquities, Naturalist's Calendar, &c.* (London, 1836), p. iii.

the 1830s, *Selborne* also provided amateur naturalists with a model and a justification for local studies. The availability of inexpensive editions from 1829 widened the audience for the work. No longer was it a polite tome for learned gentlemen, but it became a book that, with its emphasis on observation, was considered morally suitable for all social classes and age groups. Nonetheless, the Victorian science popularizer the Reverend John George Wood thought that White, who wrote as 'a naturalist to a naturalist', was too explicit for 'the library of a school or home'. Wood therefore omitted or obscured references to copulation in his edition, anxious not to offend those readers 'whose studies in Natural History are not of a rigidly scientific character'.[24] Although *Selborne* was read by an increasingly wide audience, some critics did not think that the work should be categorized as 'popular science'. One reviewer, in an unfavourable comparison of Wood's popular work with White's *Selborne*, considered that the former was read by those 'who could not properly observe a fact if they tried' and the latter by 'born observers', who sought a guide to the study of nature.[25]

The publication history of *Selborne* carries with it not only the aims and assumptions of Gilbert White, but also those of its many subsequent editors. It thus provides a remarkable window onto changing attitudes towards the natural world over more than two centuries. New editions produced by naturalists inspired by White were especially numerous in the nineteenth century, when the book also appealed to novelists such as Mary Russell Mitford and George Eliot, whose work, in contrast to Gothic or sensation fiction, portrayed the unremarkable lives of ordinary people. In the 1930s *Selborne* contributed to the concept of England held by the 'Romantic Moderns', among them Virginia Woolf.[26] Novelists such as Ford Madox Ford and John Cowper Powys could make passing reference to Gilbert White and *Selborne*, secure in the knowledge that their readers would be familiar with the author and his book. The experience of empire was also crucial in establishing the book as the embodiment of Englishness. Colonists took the book with them to far-flung places, where it served as a nostalgic reminder of 'home', even as it

---

[24] J. G. Wood (ed.), *The Natural History of Selborne* (London, 1854), pp. v–vi.

[25] 'Man and Beast Here and Hereafter', *Pall Mall Gazette* (9 February 1875), 11.

[26] Alexandra Harris, *Romantic Moderns: English Writers, Artists and the Imagination from Virginia Woolf to John Piper* (London, 2010), 171–3.

also inspired studies of indigenous faunas and floras. (The way in which readers and writers have breathed new life into the work by seeing its relevance to their own times may be seen in Appendix II, which contains a selection of responses to the book.)

The success of *Selborne* has been to make a particular mode of studying nature seem self-evident. The work continues to appear so fresh because it marks the start of a tradition; its impact and importance lay, as White's contemporaries recognized, in the 'Novelty, & Elegance, the Tenderness, & . . . *Piety* of the natural Part'.[27] White's achievement was to show that expert observation discerns and stabilizes the proper objects of study for a wider community of naturalists, as well as providing the materials from which to construct a universal natural history. The delicate and variable interconnectedness of natural phenomena, so well appreciated by White, demands attention to local circumstances. Modern environmental concerns also require this sort of knowledge and this kind of wonder. White's suggestion that any one observer should subject a small area to thorough scrutiny in order to contribute to a larger understanding of natural processes is as relevant to present issues concerning biodiversity as it was to White's physico-theological aims. The continuing appeal of *Selborne* is not confined to its role as a model for field observation and scientific writing, but extends to all the pleasures associated with losing oneself in nature.

[27] Holt-White (ed.), *Letters to Gilbert White . . . from . . . John Mulso*, 284.

NOTE ON THE TEXT

THIS edition is based on the text of the *Natural History* first published as part of *The Natural History and Antiquities of Selborne* (1789), the only edition to appear in White's lifetime. In keeping with that edition, White's footnotes appear at the bottom of the page; explanatory notes by the editor, the position of which is indicated by an asterisk, are grouped at the end of the text (pp. 271–98).

There were several typographical errors in the first edition of *Selborne*, some of which were included in the list of errata published at the end of the book. These, together with the obvious misplacement of some of White's footnotes and punctuation, have been silently corrected. The additional sections of letters 41 and 58 to Barrington, originally tacked on to the end of the volume (after the *Antiquities*), have been moved to the positions indicated by White. I have also followed Paul Foster (*Gilbert White and his Records* (London, 1988), 200, n. 9) in omitting the first sentence of letter 8 to Barrington.

Grammatical errors in the text (notably 'it's' for 'its') have been corrected but in general White's spelling, even when inconsistent, has been retained. Changes have been made only in cases where the original is puzzling or potentially misleading; for example, 'leavers' has been replaced by 'levers', and 'dosed' changed to 'dozed', in letters 21 and 27 to Daines Barrington. The eighteenth-century use of italics and capitalization has also been modernized. In accordance with current taxonomic practice, only genus and species names are italicized. The ligatures 'æ' and 'œ' have been expanded to 'ae' and 'oe'. The roman numerals used to number the letters in the first edition of *Selborne* have been replaced with Arabic ones.

Sources have been provided (with translations where appropriate) for the many quotations used by White in the text. Where White himself supplied sources, his references (in particular, those to the Bible) have been given in standard modern form.

White's 1789 quarto edition of *Selborne* included nine illustrations, only three of which (plates 2–4) related to the content of the *Natural History*. The illustrations in this edition come from the works of

Thomas Pennant, one of White's correspondents. As the text reveals, these are images with which White was familiar and on which he commented. They therefore serve to illuminate further the eighteenth-century view of the natural world that White so vividly described.

SELECT BIBLIOGRAPHY

A sample of comments on *The Natural History of Selborne* by Gilbert White's contemporaries and later readers is in Appendix II.

Editions of The Natural History of Selborne *and the Journals*

A work like *The Natural History of Selborne*, which has never been out of print and has been translated into many languages, clearly exists in a vast number of editions. A wide range of these are described in E. A. Martin, *A Bibliography of Gilbert White* (Folkestone: Dawsons, 1897; rev. edn. 1934, repr. with additions 1970). Standard electronic catalogues also provide the means for compiling an extensive bibliography. Many of the editions are freely available online through the Internet Archive and the more specialized Biodiversity Heritage Library. The following list charts in brief the increasing accessibility of *The Natural History of Selborne* in the nineteenth century; it also includes recent and easily available editions of *Selborne* and White's journals.

White, Gilbert, *The Natural History and Antiquities of Selborne, in the County of Southampton: With Engravings, and an Appendix* (London: Benjamin White and Son, 1789) [the book was, in fact, available from November 1788].

Aikin, J. (ed.), *The Works, in Natural History, of the Late Rev. Gilbert White . . . Comprising The Natural History of Selborne; The Naturalist's Calendar; and Miscellaneous Observations, Extracted from his Papers,* 2 vols. (London: J. White, 1802).

White, John (ed.), *The Natural History and Antiquities of Selborne . . . to which are Added, The Naturalist's Calendar, Observations on Various Parts of Nature, and Poems* (London: White, Cochrane, and Co., 1813).

White, Gilbert, *The Natural History of Selborne . . . to which are Added, The Naturalist's Calendar, Miscellaneous Observations, and Poems,* 2 vols. (London: C. and J. Rivington, 1825).

Jardine, William (ed.), *The Natural History of Selborne,* Constable's Miscellany, xlv (Edinburgh: Constable, 1829) [the first inexpensive, popular edition].

Bell, Thomas (ed.), *The Natural History and Antiquities of Selborne,* 2 vols. (London: Van Voorst, 1877) [with substantial notes and biographical information, including many letters published for the first time].

Johnson, Walter (ed.), *Gilbert White's Journals* (London: Routledge, 1933; repr. 1970).

Mabey, Richard (ed.), *The Natural History of Selborne* (Harmondsworth: Penguin Books, 1977).

Greenoak, F. (ed.), *The Journals of Gilbert White, 1751–93*, 3 vols. (London: Century Hutchinson, 1986–9) [the complete texts of White's 'Garden Kalendar', 1751–71, 'Calendar of Flora', 1766, and 'Naturalist's Journal', 1768–93].

Foster, Paul (ed.), *The Natural History of Selborne*, The World's Classics (Oxford: Oxford University Press, 1993).

Chatfield, June E. (ed.), *The Illustrated Natural History of Selborne* (London: Thames and Hudson, 2004).

Biographical Works and Letters

Coulson, L., *To Edify and Delight: Introducing the Sermons of Gilbert White of Selborne* (Chichester, 2011).

Foster, P., *Gilbert White and his Records: A Scientific Biography* (London, 1988).

——'The Gibraltar Collections: Gilbert White (1720–1793) and John White (1727–1780), and the Naturalist and Author Giovanni Antonio Scopoli (1723–1788)', *Archives of Natural History*, 34 (2007), 30–46.

——'White, Gilbert (1720–1793)', *Oxford Dictionary of National Biography*, Oxford University Press, 2004; online edn., Sept. 2011 <http://www.oxforddnb.com/view/article/29243>, accessed 11 Nov. 2012.

Holt-White, R., *The Life and Letters of Gilbert White of Selborne*, 2 vols. (London, 1901).

——(ed.), *The Letters to Gilbert White of Selborne from his Intimate Friend and Contemporary the Rev. John Mulso* (London, 1907; facs. edn., Cambridge, 2011).

Mabey, R., *Gilbert White: A Biography of the Naturalist and Author of The Natural History of Selborne* (London, 1986).

Critical Approaches

Bohrer, M. A., 'Tales of Locale: *The Natural History of Selborne* and *Castle Rackrent*', *Modern Philology*, 100 (2003), 393–416.

Cooper, N., 'Gilbert White and the Natural Theology of Selborne', *The Linnean*, 23/3 (2007), 29–44.

Dadswell, T., *The Selborne Pioneer: Gilbert White as Naturalist and Scientist, a Re-examination* (London, 2006).

Johnson, W., *Gilbert White: Pioneer, Poet, and Stylist* (London, 1928).

King, A. M., 'Quietism and Narrative Stillness', *Common Knowledge*, 16 (2010), 532–51.

Lipscomb, S. B., 'Introducing Gilbert White: An Exemplary Natural Historian and his Editors', *Victorian Literature and Culture*, 35 (2007), 551–67.

Maddox, L. B., 'Gilbert White and the Politics of Natural History', *Eighteenth-Century Life*, 10 (1986), 45–57.

Menely, T., 'Traveling in Place: Gilbert White's Cosmopolitan Parochialism', *Eighteenth-Century Life*, 28 (2004), 46–65.

Mullett, C. F., '*Multum in parvo*: Gilbert White of Selborne', *Journal of the History of Biology*, 2 (1969), 363–89.

Eighteenth-Century Contexts

Allen, D., *The Naturalist in Britain: A Social History* (London, 1976).

—— *Books and Naturalists* (London, 2010).

Bellanca, M. E., *Daybooks of Discovery: Nature Diaries in Britain, 1770–1870* (Charlottesville, Va., and London, 2007).

Brewer, J., *Pleasures of the Imagination: English Culture in the Eighteenth Century* (London, 1997).

Clark, W., J. Golinski, and S. Schaffer (eds.), *The Sciences in Enlightened Europe* (Chicago, 1999).

Colley, L., *Britons: Forging the Nation, 1707–1837* (New Haven and London, 1992).

Daston, L., 'The Empire of Observation, 1600–1800', in *Histories of Scientific Observation*, ed. L. Daston and E. Lunbeck (Chicago, 2011), 81–113.

Donald, D., *Picturing Animals in Britain, c.1750–1850* (New Haven and London, 2007).

Goldgar, A., *Impolite Learning: Conduct and Community in the Republic of Letters, 1680–1750* (New Haven and London, 1995).

Golinski, J., *British Weather and the Climate of Enlightenment* (Chicago, 2007).

Jardine, N., J. Secord, and E. C. Spary (eds.), *Cultures of Natural History* (Cambridge, 1996).

Noblett, W., 'Pennant and his Publisher: Benjamin White, Thomas Pennant and *Of London*', *Archives of Natural History*, 11 (1982), 61–8.

Snell, K. D. M., *Parish and Belonging: Community, Identity and Welfare in England and Wales, 1700–1950* (Cambridge, 2006).

Thomas, K., *Man and the Natural World: Changing Attitudes in England, 1500–1800* (London, 1983).

Wall, C. S., *The Prose of Things: Transformations of Description in the Eighteenth Century* (Chicago, 2006).

A CHRONOLOGY OF GILBERT WHITE

1720 (18 July) Gilbert White born in Selborne, Hampshire, eldest of eleven children of John White (1688–1758), barrister and JP for Hampshire, and Anne, *née* Holt (1693–1739).

1723 Black Act makes theft and poaching capital offences.

1727 George II becomes king.

1728 White's grandfather, Gilbert White, vicar of Selborne, dies.

1729 White's parents move from Surrey to Selborne and settle at The Wakes, which remains White's home for the rest of his life. Treaty of Seville gives control of Gibraltar to Britain.

1730 White celebrates his 10th birthday by planting an oak and an ash tree at The Wakes.

1733 Sent to study with the Reverend Thomas Warton at Basingstoke. Use of Latin in parish registers prohibited by law.

1736 Earliest surviving natural history observations, made on a holiday in Rutland.

1739 (17 December) Death of his mother. England at war with Spain (until 1748).

1740 Begins studies at Oriel College, Oxford.

1741 (21 September) Observes a fall of gossamer.

1743 Graduates from Oxford, and is presented by Alexander Pope with a copy of his translation of Homer's *Iliad*.

1744 Elected Fellow of Oriel College, Oxford.

1745 Jacobite rising in Scotland.

1746 Spends six months in Ely, settling the estate of a relative.

1747 Made deacon at Christ Church, Oxford, and appointed curate at Swarraton, Hampshire. Suffers an attack of smallpox. Buys Philip Miller's *Gardener's Dictionary*.

1751 Serves as curate at Selborne for five months. Makes improvements to the garden at The Wakes.

1751–67 Keeps detailed records in his 'Garden Kalendar'.

1752 Dean of Oriel College, Oxford. (8 April) Appointed junior proctor in the University of Oxford for a year. On 3 September the date moves forward by eleven days with the adoption of the Gregorian calendar in England, making this 14 September.

1753 New year begins on 1 January, not 25 March, as part of calendar reform. White appointed curate at Durley, Hampshire (for eighteen months). Swedish naturalist Carl Linnaeus publishes *Species plantarum*.

1755 Lisbon earthquake; 90,000 killed.

1755–6 Curate at West Dean, Wiltshire.

1755–7 Curate at Newton Valence, Hampshire.

1756–7 Curate at Selborne.

1756–63 Seven Years War.

1757 Unsuccessful as candidate for Provost of Oriel College, Oxford. Elected perpetual curate (non-resident) at Moreton Pinkney, Northamptonshire.

1758 (29 September) Death of his father.

1758–9 Curate at Selborne.

1759 British Museum in London opens to the public.

1760 Tour to Rutland. George III becomes king. Enclosure Acts increasingly replace common lands with private ownership.

1761 Builds fruit-wall and ha-ha at The Wakes.

1761–84 Curate at Farringdon, Hampshire.

1763 Inherits The Wakes, where he has been living since 1729.

1765 Purchases William Hudson's *Flora Anglica* (1762).

1766 Compiles 'Flora Selborniensis'. Publication of the first edition of *British Zoology* by the Welsh naturalist Thomas Pennant.

1767 Meets Joseph Banks and Daniel Solander in London, before their voyage with Captain James Cook. Begins correspondence with Pennant, providing him with local information.

1768 First edition of *Encyclopedia Britannica* published in Edinburgh.

1768–72 Captain James Cook's circumnavigation of the globe.

1768–93 Records regular observations in his 'Naturalist's Journal'.

1769 Begins correspondence with Daines Barrington. Reads Giovanni Antonio Scopoli's monograph on the birds of Carniola in Slovenia. Encourages his brother John White, chaplain at Gibraltar, to pursue natural history.

1770 Barrington encourages White to write a natural history of Selborne. Captain James Cook lands in Australia at Botany Bay.

1772 John White returns from Gibraltar and spends the winter at Selborne, discussing natural history with Gilbert.

1774 (10 February) White's account of the house martin is read at a meeting of the Royal Society of London.

1775 (16 March) White's accounts of the house swallow, swift, and sand martin are read at a meeting of the Royal Society of London.

1775–83 American War of Independence.

1776 Samuel H. Grimm makes twelve watercolour drawings of Selborne for White's projected book.

1777 White's brother Thomas White is elected Fellow of the Royal Society of London.

1780 Inherits Timothy the tortoise from his aunt Rebecca Snooke. (2–8 June) Gordon Riots in London. Death of White's brother John White.

1781 Barbara White, John White's widow, comes to live at The Wakes. William Herschel discovers the *Georgium Sidus* (the planet Uranus).

1782 Establishment of outdoor poor relief in England.

1783 The Montgolfier brothers launch the first hot-air balloon in France. Excessively hot July caused by a volcanic eruption in Iceland. The last public execution carried out at Tyburn in London.

1784 White observes Jean-Pierre Blanchard's balloon pass over Selborne. The botanist James Edward Smith purchases Linnaeus's collections.

1784–93 Curate at Selborne (the vicar was non-resident).

1785 Publication of the first edition of the *Daily Universal Register* (renamed *The Times* in 1788).

1788 (1 November) Publication of *The Natural History and Antiquities of Selborne*, dated 1789 (books published in the closing months of the year could carry the following year's publication date). The first convicts and settlers arrive in New South Wales. Founding of the Linnean Society of London.

1789 (14 July) French Revolution begins.

1791 (21 September) Death of his friend and correspondent the Reverend John Mulso.

1793 (21 January) Execution of Louis XVI; (8 February) England declares war on France. (15 June) White makes the last entry in his 'Naturalist's Journal'. (26 June) White dies at The Wakes, and (1 July) is buried in Selborne churchyard.

THE NATURAL HISTORY
OF SELBORNE

In a series of letters addressed to

THOMAS PENNANT, ESQ.
and
THE HON. DAINES BARRINGTON

ego Apis Matinae
More modoque
Grata carpentis . . . per laborem
Plurimum . . .

(Horace)*

Omnia benè describere, quae in hoc mundo, a Deo facta, aut
Naturae creatae viribus elaborata fuerunt, opus est non unius
hominis, nec unius aevi. Hinc Faunae & Florae utilissimae;
hinc Monographi praestantissimi.

(Scopoli *Ann. Hist. Nat.*)

τρηχεῖ, ἀλλ᾽ ἀγαθὴ κουροτρόφος· οὔ τι ἐγώ γε
ἧς γαίης δύναμαι γλυκερώτερον ἄλλο ἰδέσθαι.
(Homer, *Odyssey*)

Tota denique nostra illa aspera, & montuosa, & fidelis, &
simplex, & fautrix suorum regio.

(Cicero, *Oratio pro Cn. Plancio*)

ADVERTISEMENT

The author of the following letters takes the liberty, with all proper deference, of laying before the public his idea of *parochial history*, which, he thinks, ought to consist of natural productions and occurrences as well as antiquities. He is also of opinion that if stationary men* would pay some attention to the districts on which they reside, and would publish their thoughts respecting the objects that surround them, from such materials might be drawn the most complete county-histories, which are still wanting in several parts of this kingdom, and in particular in the county of Southampton.*

And here he seizes the first opportunity, though a late one, of returning his most grateful acknowledgments to the reverend the President and the reverend and worthy the Fellows of Magdalen College in the university of Oxford, for their liberal behaviour in permitting their archives to be searched by a member of their own society, so far as the evidences therein contained might respect the parish and priory of Selborne. To that gentleman also, and his assistant, whose labours and attention could only be equalled by the very kind manner in which they were bestowed, many and great obligations are also due.*

Of the authenticity of the documents above-mentioned there can be no doubt, since they consist of the identical deeds and records that were removed to the College from the Priory at the time of its dissolution; and, being carefully copied on the spot, may be depended on as genuine; and, never having been made public before, may gratify the curiosity of the antiquary, as well as establish the credit of the history.

If the writer should at all appear to have induced any of his readers to pay a more ready attention to the wonders of the Creation, too frequently overlooked as common occurrences; or if he should by any means, through his researches, have lent an helping hand towards the enlargement of the boundaries of historical and topographical knowledge; or if he should have thrown some small light upon ancient customs and manners, and especially on those that were monastic; his purpose will be fully answered. But if he should not have been successful in any of these his intentions, yet there remains this consolation behind—that these his pursuits, by keeping the body and mind employed, have, under Providence, contributed to much health and cheerfulness of spirits, even to old age: and, what still adds to his happiness, have led him to the knowledge of a circle of gentlemen whose intelligent communications, as they have afforded him much pleasing information, so, could he flatter himself with a continuation of them, would they ever be deemed a matter of singular satisfaction and improvement.

GIL. WHITE

Selborne,
January 1st, 1788

LETTERS

TO

THOMAS PENNANT, ESQUIRE

LETTER 1

The parish of Selborne lies in the extreme eastern corner of the county of Hampshire, bordering on the county of Sussex, and not far from the county of Surrey; is about fifty miles south-west of London, in latitude 51, and near midway between the towns of Alton and Petersfield. Being very large and extensive it abuts on twelve parishes, two of which are in Sussex, *viz.* Trotton and Rogate. If you begin from the south and proceed westward the adjacent parishes are Emshot, Newton Valence, Faringdon, Harteley Mauduit, Great Ward le ham, Kingsley, Hedleigh, Bramshot, Trotton, Rogate, Lysse, and Greatham. The soils of this district are almost as various and diversified as the views and aspects. The high part to the south-west consists of a vast hill of chalk, rising three hundred feet above the village; and is divided into a sheep down, the high wood, and a long hanging wood called the Hanger.* The covert of this eminence is altogether beech, the most lovely of all forest trees, whether we consider its smooth rind or bark, its glossy foliage, or graceful pendulous boughs. The down, or sheep-walk, is a pleasing park-like spot, of about one mile by half that space, jutting out on the verge of the hill-country, where it begins to break down into the plains, and commanding a very engaging view, being an assemblage of hill, dale, wood-lands, heath, and water. The prospect is bounded to the south-east and east by the vast range of mountains called the Sussex Downs, by Guild-down near Guildford, and by the Downs round Dorking, and Ryegate in Surrey, to the north-east, which altogether, with the country beyond Alton and Farnham, form a noble and extensive outline.

At the foot of this hill, one stage or step from the uplands, lies the village, which consists of one single straggling street, three quarters of a mile in length, in a sheltered vale, and running parallel with the Hanger. The houses are divided from the hill by a vein of stiff clay (good wheat-land), yet stand on a rock of white stone, little in appearance removed from chalk; but seems so far from being calcarious, that it endures extreme heat. Yet that the freestone* still preserves somewhat that is analogous to chalk, is plain from the beeches which descend as low as those rocks extend, and no farther, and thrive as well on them, where the ground is steep, as on the chalks.

The cart-way of the village divides, in a remarkable manner, two very incongruous soils. To the south-west is a rank clay, that requires the labour of years to render it mellow; while the gardens to the north-east, and small enclosures behind, consist of a warm, forward, crumbling mould, called black malm, which seems highly saturated with vegetable and animal manure; and these may perhaps have been the original site of the town; while the woods and coverts* might extend down to the opposite bank.

At each end of the village, which runs from south-east to north-west, arises a small rivulet: that at the north-west end frequently fails; but the other is a fine perennial spring, little influenced by drought or wet seasons, called Well-head.[1] This breaks out of some high grounds joining to Nore Hill, a noble chalk promontory, remarkable for sending forth two streams into two different seas. The one to the south becomes a branch of the Arun, running to Arundel, and so falling into the British channel: the other to the north. The Selborne stream makes one branch of the Wey; and, meeting the Black-down stream at Hedleigh, and the Alton and Farnham stream at Tilford-bridge, swells into a considerable river, navigable at Godalming; from whence it passes to Guildford, and so into the Thames at Weybridge; and thus at the Nore into the German ocean.*

Our wells, at an average, run to about sixty-three feet, and when sunk to that depth seldom fail; but produce a fine limpid water, soft to the taste, and much commended by those who drink the pure element, but which does not lather well with soap.

To the north-west, north and east of the village, is a range of fair enclosures, consisting of what is called a white malm, a sort of rotten or rubble stone, which, when turned up to the frost and rain, moulders to pieces, and becomes manure to itself.[2]

Still on to the north-east, and a step lower, is a kind of white land, neither chalk nor clay, neither fit for pasture nor for the plough, yet kindly for hops, which root deep into the freestone, and have their poles and wood for charcoal growing just at hand. This white soil produces the brightest hops.

---

[1] This spring produced, September 14, 1781, after a severe hot summer, and a preceding dry spring and winter, nine gallons of water in a minute, which is five hundred and forty in an hour, and twelve thousand nine hundred and sixty, or two hundred and sixteen hogsheads, in twenty-four hours, or one natural day. At this time many of the wells failed, and all the ponds in the vales were dry.

[2] This soil produces good wheat and clover.

As the parish still inclines down towards Wolmer-forest,* at the juncture of the clays and sand the soil becomes a wet, sandy loam, remarkable for timber, and infamous for roads. The oaks of Temple and Blackmoor stand high in the estimation of purveyors, and have furnished much naval timber; while the trees on the freestone grow large, but are what workmen call shakey,* and so brittle as often to fall to pieces in sawing. Beyond the sandy loam the soil becomes an hungry lean sand, till it mingles with the forest; and will produce little without the assistance of lime and turnips.*

LETTER 2

In the court of Norton farm house, a manor farm to the northwest of the village, on the white malms, stood within these twenty years a broad-leaved elm, or wych hazel, *ulmus folio latissimo scabro* of Ray, which, though it had lost a considerable leading bough in the great storm in the year 1703,* equal to a moderate tree, yet, when felled, contained eight loads of timber; and, being too bulky for a carriage, was sawn off at seven feet above the butt, where it measured near eight feet in the diameter. This elm I mention to shew to what a bulk planted elms may attain; as this tree must certainly have been such from its situation.

In the centre of the village, and near the church, is a square piece of ground surrounded by houses, and vulgarly called the Plestor.[1]* In the midst of this spot stood, in old times, a vast oak, with a short squat body, and huge horizontal arms extending almost to the extremity of the area. This venerable tree, surrounded with stone steps, and seats above them, was the delight of old and young, and a place of much resort in summer evenings; where the former sat in grave debate, while the latter frolicked and danced before them. Long might it have stood, had not the amazing tempest in 1703 overturned it at once, to the infinite regret of the inhabitants, and the vicar,* who bestowed several pounds in setting it in its place again: but all his care could not avail; the tree sprouted for a time, then withered and died. This oak I mention to shew to what a bulk planted oaks also may arrive: and planted this tree must certainly have been, as will appear from

---

[1] Vide the plate in the *Antiquities*.

what will be said farther concerning this area, when we enter on the antiquities of Selborne.

On the Blackmoor estate there is a small wood called Losel's, of a few acres, that was lately furnished with a set of oaks of a peculiar growth and great value; they were tall and taper like firs, but standing near together had very small heads, only a little brush without any large limbs. About twenty years ago the bridge at the Toy, near Hampton Court, being much decayed, some trees were wanted for the repairs* that were fifty feet long without bough, and would measure twelve inches diameter at the little end. Twenty such trees did a purveyor find in this little wood, with this advantage, that many of them answered the description at sixty feet. These trees were sold for twenty pounds apiece.

In the centre of this grove there stood an oak, which, though shapely and tall on the whole, bulged out into a large excrescence about the middle of the stem. On this a pair of ravens had fixed their residence for such a series of years, that the oak was distinguished by the title of the Raven-tree. Many were the attempts of the neighbouring youths to get at this eyry:* the difficulty whetted their inclinations, and each was ambitious of surmounting the arduous task. But, when they arrived at the swelling, it jutted out so in their way, and was so far beyond their grasp, that the most daring lads were awed, and acknowledged the undertaking to be too hazardous. So the ravens built on, nest upon nest, in perfect security, till the fatal day arrived in which the wood was to be levelled. It was in the month of February, when those birds usually sit. The saw was applied to the butt, the wedges were inserted into the opening, the woods echoed to the heavy blows of the beetle or mallet, the tree nodded to its fall; but still the dam sat on. At last, when it gave way, the bird was flung from her nest; and, though her parental affection deserved a better fate, was whipped down by the twigs, which brought her dead to the ground.

LETTER 3

The fossil-shells of this district, and sorts of stone, such as have fallen within my observation, must not be passed over in silence. And first I must mention, as a great curiosity, a specimen that was plowed up in the chalky fields, near the side of the Down, and given to me for the singularity of its appearance, which, to an incurious eye, seems like a petrified fish of about four inches long, the cardo* passing for an head and mouth. It is in reality a bivalve of the Linnaean genus of *Mytilus*, and the species of *Crista Galli*; called by Lister, *Rastellum*; by Rumphius, *Ostreum plicatum minus*; by D'Argenville, *Auris Porci*, s. *Crista Galli*; and by those who make collections cock's comb. Though I applied to several such in London, I never could meet with an entire specimen; nor could I ever find in books any engraving from a perfect one. In the superb museum at Leicester-house permission was given me to examine for this article;* and, though I was disappointed as to the fossil, I was highly gratified with the sight of several of the shells themselves in high preservation. This bivalve is only known to inhabit the Indian ocean, where it fixes itself to a zoophyte, known by the name *Gorgonia*.* The curious foldings of the suture the one into the other, the alternate flutings or grooves, and the curved form of my specimen being much easier expressed by the pencil than by words, I have caused it to be drawn and engraved.*

*Cornua Ammonis** are very common about this village. As we were cutting an inclining path up the Hanger,* the labourers found them frequently on that steep, just under the soil, in the chalk, and of a considerable size. In the lane above Well-head, in the way to Emshot, they abound in the bank in a darkish sort of marl; and are usually very small and soft: but in Clay's Pond, a little farther on, at the end of the pit, where the soil is dug out for manure, I have occasionally observed them of large dimensions, perhaps fourteen or sixteen inches in diameter. But as these did not consist of firm stone, but were formed of a kind of *terra lapidosa*, or hardened clay, as soon as they were exposed to the rains and frost they mouldered away. These seemed as if they were a very recent production.* In the chalk-pit, at the north-west end of the Hanger, large *nautili** are sometimes observed.

In the very thickest strata of our freestone, and at considerable depths, well-diggers often find large scallops or pectines, having both shells deeply striated, and ridged and furrowed alternately. They are highly impregnated with, if not wholly composed of, the stone of the quarry.

LETTER 4

As in a former letter the freestone of this place has been only mentioned incidentally, I shall here become more particular.

This stone is in great request for hearth-stones, and the beds of ovens: and in lining of lime-kilns it turns to good account; for the workmen use sandy loam instead of mortar; the sand of which fluxes,[1] and runs by the intense heat, and so cases over the whole face of the kiln with a strong vitrified coat like glass, that it is well preserved from injuries of weather, and endures thirty or forty years. When chiseled smooth, it makes elegant fronts for houses, equal in colour and grain to the Bath stone; and superior in one respect, that, when seasoned, it does not scale. Decent chimney-pieces are worked from it of much closer and finer grain than Portland;* and rooms are floored with it; but it proves rather too soft for this purpose. It is a freestone, cutting in all directions; yet has something of a grain parallel with the horizon, and therefore should not be surbedded, but laid in the same position that it grows in the quarry.[2] On the ground abroad this firestone will not succeed for pavements, because, probably some degree of saltness prevailing within it, the rain tears the slabs to pieces.[3] Though this stone is too hard to be acted on by vinegar; yet both the white part, and even the blue rag, ferments strongly in mineral acids. Though the white stone will not bear wet, yet in every quarry at intervals there are thin strata of blue rag, which resist rain and frost; and are excellent for pitching of stables, paths and courts, and for building of dry walls against banks; a valuable species of fencing, much in use in this village, and for mending of roads. This rag is rugged and stubborn, and will not hew to a smooth face; but is very durable: yet, as these strata are shallow and lie deep, large quantities cannot be procured but at considerable expense. Among the blue rags turn up some blocks tinged with a stain of yellow or rust

[1] There may probably be also in the chalk itself that is burnt for lime a proportion of sand: for few chalks are so pure as to have none.

[2] To surbed stone is to set it edgewise, contrary to the posture it had in the quarry, says Dr Plot. *Oxfordsh.* p. 77.* But surbedding does not succeed in our dry walls; neither do we use it so in ovens, though he says it is best for Teynton stone.

[3] 'Firestone is full of salts, and has no sulphur: must be close grained, and have no interstices. Nothing supports fire like salts; saltstone perishes exposed to wet and frost.' Plot's *Staff*, p. 152.*

colour, which seem to be nearly as lasting as the blue; and every now and then balls of a friable substance, like rust of iron, called rust balls.

In Wolmer Forest I see but one sort of stone, called by the work-men sand, or forest-stone. This is generally of the colour of rusty iron, and might probably be worked as iron ore; is very hard and heavy, and of a firm, compact texture, and composed of a small roundish crystal-line grit, cemented together by a brown, terrene,* ferruginous matter; will not cut without difficulty, nor easily strike fire with steel. Being often found in broad flat pieces, it makes good pavement for paths about houses, never becoming slippery in frost or rain; is excellent for dry walls, and is sometimes used in buildings. In many parts of that waste it lies scattered on the surface of the ground; but is dug on Weaver's Down, a vast hill on the eastern verge of that forest, where the pits are shallow, and the stratum thin. This stone is imperishable.

From a notion of rendering their work the more elegant, and giving it a finish, masons chip this stone into small fragments about the size of the head of a large nail; and then stick the pieces into the wet mortar along the joints of their freestone walls: this embellishment carries an odd appearance, and has occasioned strangers sometimes to ask us pleasantly, 'whether we fastened our walls together with tenpenny nails.'

LETTER 5

Among the singularities of this place the two rocky hollow lanes, the one to Alton, and the other to the forest, deserve our attention. These roads, running through the malm lands, are, by the traffick of ages, and the fretting of water, worn down through the first stratum of our freestone, and partly through the second; so that they look more like water-courses than roads; and are bedded with naked rag for fur-longs* together. In many places they are reduced sixteen or eighteen feet beneath the level of the fields; and after floods, and in frosts, exhibit very grotesque and wild appearances, from the tangled roots that are twisted among the strata, and from the torrents rushing down their broken sides; and especially when those cascades are frozen into icicles, hanging in all the fanciful shapes of frost-work. These rugged gloomy scenes affright the ladies when they peep down into them from the paths above, and make timid horsemen shudder while they ride along them; but delight the naturalist with their various botany, and particularly with their curious filices* with which they abound.

The manor of Selborne, was it strictly looked after, with all its kindly aspects, and all its sloping coverts, would swarm with game; even now hares, partridges, and pheasants abound; and in old days woodcocks were as plentiful. There are few quails, because they more affect* open fields than enclosures; after harvest some few land-rails are seen.

The parish of Selborne, by taking in so much of the forest, is a vast district. Those who tread the bounds* are employed part of three days in the business, and are of opinion that the outline, in all its curves and indentings, does not comprise less than thirty miles.

The village stands in a sheltered spot, secured by the Hanger from the strong westerly winds. The air is soft, but rather moist from the effluvia of so many trees; yet perfectly healthy and free from agues.

The quantity of rain that falls on it is very considerable, as may be supposed in so woody and mountainous a district. As my experience in measuring the water is but of short date, I am not qualified to give the mean quantity.[1] I only know that

	Inch.	Hund.
From May 1, 1779, to the end of the year there fell	28	37!
From Jan. 1, 1780, to Jan. 1, 1781	27	32
From Jan. 1, 1781, to Jan. 1, 1782	30	71
From Jan. 1, 1782, to Jan. 1, 1783	50	26!
From Jan. 1, 1783, to Jan. 1, 1784	33	71
From Jan. 1, 1784, to Jan. 1, 1785	33	80
From Jan. 1, 1785, to Jan. 1, 1786	31	55
From Jan. 1, 1786, to Jan. 1, 1787	39	57

The village of Selborne, and large hamlet of Oakhanger, with the single farms, and many scattered houses along the verge of the forest, contain upwards of six hundred and seventy inhabitants.[2]

[1] A very intelligent gentleman* assures me (and he speaks from upwards of forty years experience) that the mean rain of any place cannot be ascertained till a person has measured it for a very long period. 'If I had only measured the rain,' says he, 'for the four first years, from 1740 to 1743, I should have said the mean rain at Lyndon was 16 1-hf. inch for the year; if from 1740 to 1750, 18 1-hf. inches. The mean rain before 1763 was 20 1-qr. from 1763 and since 25 1-hf. from 1770 to 1780, 26. If only 1773, 1774 and 1775, had been measured, Lyndon mean rain would have been called 32 inches.'

[2] [For White's footnote on the population of Selborne, see Appendix I.]

We abound with poor; many of whom are sober and industrious, and live comfortably in good stone or brick cottages, which are glazed, and have chambers above stairs: mud buildings we have none. Besides the employment from husbandry, the men work in hop gardens, of which we have many; and fell and bark timber.* In the spring and summer the women weed the corn; and enjoy a second harvest in September by hop picking. Formerly, in the dead months they availed themselves greatly by spinning wool, for making of barragons, a genteel corded stuff, much in vogue at that time for summer wear; and chiefly manufactured at Alton, a neighbouring town, by some of the people called Quakers: but from circumstances this trade is at an end.[1] The inhabitants enjoy a good share of health and longevity; and the parish swarms with children.

LETTER 6

Should I omit to describe with some exactness the forest of Wolmer, of which three fifths perhaps lie in this parish, my account of Selborne would be very imperfect, as it is a district abounding with many curious productions, both animal and vegetable; and has often afforded me much entertainment both as a sportsman and as a naturalist.

The royal forest of Wolmer is a tract of land of about seven miles in length, by two and a half in breadth, running nearly from North to South, and is abutted on, to begin to the South, and so to proceed eastward, by the parishes of Greatham, Lysse, Rogate, and Trotton, in the county of Sussex; by Bramshot, Hedleigh, and Kingsley. This royalty consists entirely of sand covered with heath and fern; but is somewhat diversified with hills and dales, without having one standing tree in the whole extent. In the bottoms, where the waters stagnate, are many bogs, which formerly abounded with subterraneous trees; though Dr Plot says positively,[2] that 'there never were any fallen trees hidden in the mosses of the southern counties.'* But he was mistaken: for I myself have seen cottages on the verge of this wild district, whose timbers consisted of a black hard wood, looking like

---

[1] Since the passage above was written, I am happy in being able to say that the spinning employment is a little revived, to the no small comfort of the industrious housewife.

[2] See his *Hist. of Staffordshire*.

oak, which the owners assured me they procured from the bogs by probing the soil with spits, or some such instruments: but the peat is so much cut out, and the moors have been so well examined, that none has been found of late.[1] Besides the oak, I have also been shewn pieces of fossil-wood of a paler colour, and softer nature, which the inhabitants called fir: but, upon a nice examination, and trial by fire, I could discover nothing resinous in them; and therefore rather suppose that they were parts of a willow or alder, or some such aquatic tree.

This lonely domain is a very agreeable haunt for many sorts of wild fowls, which not only frequent it in the winter, but breed there in the summer; such as lapwings, snipes, wild-ducks, and, as I have discovered within these few years, teals. Partridges in vast plenty are bred in good seasons on the verge of this forest, into which they love to make excursions: and in particular, in the dry summer of 1740 and 1741, and some years after, they swarmed to such a degree that parties of unreasonable sportsmen killed twenty and sometimes thirty brace in a day.

But there was a nobler species of game in this forest, now extinct, which I have heard old people say abounded much before shooting flying* became so common, and that was the heath-cock, black game, or grouse. When I was a little boy I recollect one coming now and then to my father's table. The last pack remembered was killed about thirty-five years ago; and within these ten years one solitary grey hen was sprung by some beagles in beating for a hare. The sportsmen cried out, 'A hen pheasant;' but a gentleman present, who had

---

[1] Old people have assured me, that on a winter's morning they have discovered these trees, in the bogs, by the hoar frost, which lay longer over the space where they were concealed, than on the surrounding morass. Nor does this seem to be a fanciful notion, but consistent with true philosophy. Dr Hales saith, 'That the warmth of the earth, at some depth under ground, has an influence in promoting a thaw, as well as the change of the weather from a freezing to a thawing state, is manifest, from this observation, *viz.* Nov. 29, 1731, a little snow having fallen in the night, it was, by eleven the next morning, mostly melted away on the surface of the earth, except in several places in Bushy-park, where there were drains dug and covered with earth, on which the snow continued to lie, whether those drains were full of water or dry; as also where elm-pipes lay under ground: a plain proof this, that those drains intercepted the warmth of the earth from ascending from greater depths below them: for the snow lay where the drain had more than four feet depth of earth over it. It continued also to lie on thatch, tiles, and the tops of walls.' See Hales's *Haemastatics:** p. 360. *Quere*, Might not such observations be reduced to domestic use, by promoting the discovery of old obliterated drains and wells about houses; and in Roman stations and camps lead to the finding of pavements, baths and graves, and other hidden relics of curious antiquity?

often seen grouse in the north of England, assured me that it was a greyhen.

Nor does the loss of our black game prove the only gap in the *Fauna Selborniensis*; for another beautiful link in the chain of beings is wanting,* I mean the red deer, which toward the beginning of this century amounted to about five hundred head, and made a stately appearance. There is an old keeper, now alive, named Adams, whose great grandfather (mentioned in a perambulation taken in 1635), grandfather, father and self, enjoyed the head keepership of Wolmer forest in succession for more than an hundred years. This person assures me, that his father has often told him, that Queen Anne, as she was journeying on the Portsmouth road, did not think the forest of Wolmer beneath her royal regard. For she came out of the great road at Lippock, which is just by, and, reposing herself on a bank smoothed for that purpose, lying about half a mile to the east of Wolmer-pond, and still called Queen's-bank, saw with great complacency and satisfaction the whole herd of red deer brought by the keepers along the vale before her, consisting then of about five hundred head. A sight this worthy the attention of the greatest sovereign! But he farther adds that, by means of the Waltham blacks,* or, to use his own expression, as soon as they began blacking, they were reduced to about fifty head, and so continued decreasing till the time of the late Duke of Cumberland. It is now more than thirty years ago that his highness sent down an huntsman, and six yeomen-prickers,* in scarlet jackets laced with gold, attended by the stag-hounds; ordering them to take every deer in this forest alive, and to convey them in carts to Windsor. In the course of the summer they caught every stag, some of which shewed extraordinary diversion: but, in the following winter, when the hinds were also carried off, such fine chases were exhibited as served the country people for matter of talk and wonder for years afterwards. I saw myself one of the yeomen-prickers single out a stag from the herd, and must confess that it was the most curious feat of activity I ever beheld, superior to any thing in Mr Astley's riding-school.* The exertions made by the horse and deer much exceeded all my expectations; though the former greatly excelled the latter in speed. When the devoted deer* was separated from his companions, they gave him, by their watches, law, as they called it, for twenty minutes; when, sounding their horns, the stop-dogs were permitted to pursue, and a most gallant scene ensued.

LETTER 7

Though large herds of deer do much harm to the neighbourhood, yet the injury to the morals of the people is of more moment than the loss of their crops. The temptation is irresistible; for most men are sportsmen by constitution: and there is such an inherent spirit for hunting in human nature, as scarce any inhibitions can restrain. Hence, towards the beginning of this century all this country was wild about deer-stealing. Unless he was a hunter, as they affected to call themselves, no young person was allowed to be possessed of manhood or gallantry. The Waltham blacks at length committed such enormities, that government was forced to interfere with that severe and sanguinary act called the *black act*,[1] which now comprehends more felonies than any law that ever was framed before.* And, therefore, a late bishop of Winchester, when urged to re-stock Waltham-chase,[2] refused, from a motive worthy of a prelate, replying that 'it had done mischief enough already.'

Our old race of deer-stealers are hardly extinct yet: it was but a little while ago that, over their ale, they used to recount the exploits of their youth; such as watching the pregnant hind to her lair, and, when the calf was dropped, paring its feet with a penknife to the quick to prevent its escape, till it was large and fat enough to be killed; the shooting at one of their neighbours with a bullet in a turnip-field by moonshine, mistaking him for a deer; and the losing a dog in the following extraordinary manner:—Some fellows, suspecting that a calf new-fallen was deposited in a certain spot of thick fern, went, with a lurcher,* to surprise it; when the parent-hind rushed out of the brake, and, taking a vast spring with all her feet close together, pitched upon the neck of the dog, and broke it short in two.

Another temptation to idleness and sporting was a number of rabbits, which possessed all the hillocks and dry places: but these being inconvenient to the huntsmen, on account of their burrows, when they came to take away the deer, they permitted the country people to destroy them all.

Such forests and wastes, when their allurements to irregularities are removed, are of considerable service to neighbourhoods that

---

[1] Statute 9 Geo. I. c. 22.
[2] This chase remains un-stocked to this day; the bishop was Dr Hoadly.*

verge upon them, by furnishing them with peat and turf for their firing; with fuel for the burning their lime; and with ashes for their grasses; and by maintaining their geese and their stock of young cattle at little or no expense.

The manor-farm of the parish of Greatham has an admitted claim, I see, (by an old record taken from the Tower of London) of turning all live stock on the forest, at proper seasons, *bidentibus exceptis* [except sheep].[1] The reason, I presume, why sheep[2] are excluded, is, because, being such close grazers, they would pick out all the finest grasses, and hinder the deer from thriving.

Though (by statute 4 and 5 W. and Mary c. 23) 'to burn on any waste, between Candlemas and Midsummer,* any grig, ling, heath and furze, goss or fern, is punishable with whipping and confinement in the house of correction;' yet, in this forest, about March or April, according to the dryness of the season, such vast heath-fires are lighted up, that they often get to a masterless head, and, catching the hedges, have sometimes been communicated to the underwoods, woods, and coppices, where great damage has ensued. The plea for these burnings is, that, when the old coat of heath, &c. is consumed, young will sprout up, and afford much tender brouze* for cattle; but, where there is large old furze, the fire, following the roots, consumes the very ground; so that for hundreds of acres nothing is to be seen but smother and desolation, the whole circuit round looking like the cinders of a volcano; and, the soil being quite exhausted; no traces of vegetation are to be found for years. These conflagrations, as they take place usually with a north-east or east wind, much annoy this village with their smoke, and often alarm the country; and, once in particular, I remember that a gentleman, who lives beyond Andover,* coming to my house, when he got on the downs between that town and Winchester, at twenty-five miles distance, was surprised much with smoke and a hot smell of fire; and concluded that Alresford was in flames; but, when he came to that town, he then had apprehensions for the next village, and so on to the end of his journey.

On two of the most conspicuous eminences of this forest stand two arbours or bowers, made of the boughs of oaks; the one called

---

[1] For this privilege the owner of that estate used to pay to the king annually seven bushels of oats.

[2] In the Holt, where a full stock of fallow-deer has been kept up till lately, no sheep are admitted to this day.

Waldon-lodge, the other Brimstone-lodge: these the keepers renew annually on the feast of St Barnabas,* taking the old materials for a perquisite. The farm called Blackmoor, in this parish, is obliged to find the posts and brush-wood for the former; while the farms at Greatham, in rotation, furnish for the latter; and are all enjoined to cut and deliver the materials at the spot. This custom I mention, because I look upon it to be of very remote antiquity.

LETTER 8

On the verge of the forest, as it is now circumscribed, are three con-siderable lakes, two in Oakhanger, of which I have nothing particular to say; and one called Bin's, or Bean's pond, which is worthy the attention of a naturalist or a sportsman. For, being crowded at the upper end with willows, and with the *carex cespitosa*,[1]* it affords such a safe and pleasing shelter to wild ducks, teals, snipes, &c. that they breed there. In the winter this covert is also frequented by foxes, and sometimes by pheasants; and the bogs produce many curious plants. (For which consult letter 42 to Mr Barrington.)

By a perambulation of Wolmer forest and the Holt, made in 1635, and in the eleventh year of Charles the First (which now lies before me), it appears that the limits of the former are much circumscribed.[2] For, to say nothing of the farther side, with which I am not so well acquainted, the bounds on this side, in old times, came into Binswood; and extended to the ditch of Ward le ham-park, in which stands the curious mount called King John's Hill, and Lodge Hill; and to the verge of Hartley Mauduit, called Mauduit-hatch; comprehending also Short-heath, Oakhanger, and Oakwoods; a large district, now private property, though once belonging to the royal domain.

It is remarkable that the term purlieu* is never once mentioned in this long roll of parchment. It contains, besides the perambulation, a rough estimate of the value of the timbers, which were considerable, growing at that time in the district of the Holt; and enumerates the

---

[1] I mean that sort which, rising into tall hassocks,* is called by the foresters torrets; a corruption, I suppose of turrets.

[2] In the beginning of the summer 1787 the royal forests of Wolmer and Holt were measured by persons sent down by government.

officers, superior and inferior, of those joint forests, for the time being, and their ostensible* fees and perquisites. In those days, as at present, there were hardly any trees in Wolmer forest.

Within the present limits of the forest are three considerable lakes, Hogmer, Cranmer, and Wolmer; all of which are stocked with carp, tench, eels, and perch: but the fish do not thrive well, because the water is hungry, and the bottoms are a naked sand.*

A circumstance respecting these ponds, though by no means peculiar to them, I cannot pass over in silence; and that is, that instinct by which in summer all the kine, whether oxen, cows, calves, or heifers, retire constantly to the water during the hotter hours; where, being more exempt from flies, and inhaling the coolness of that element, some belly deep, and some only to mid-leg, they ruminate and solace themselves from about ten in the morning till four in the afternoon, and then return to their feeding. During this great proportion of the day they drop much dung, in which insects nestle; and so supply food for the fish, which would be poorly subsisted but from this contingency. Thus Nature, who is a great economist, converts the recreation of one animal to the support of another! Thomson, who was a nice* observer of natural occurrences, did not let this pleasing circumstance escape him. He says, in his 'Summer',

> A various group the herds and flocks compose:
> . . . . on the grassy bank
> Some ruminating lie; while others stand
> Half in the flood, and, often bending, sip
> The circling surface.

[James Thomson, *The Seasons* (1730), lines 485–9]

Wolmer-pond, so called, I suppose, for eminence sake, is a vast lake for this part of the world, containing, in its whole circumference, 2646 yards, or very near a mile and an half. The length of the north-west and opposite side is about 704 yards, and the breadth of the south-west end about 456 yards. This measurement, which I caused to be made with good exactness, gives an area of about sixty-six acres, exclusive of a large irregular arm at the north-east corner, which we did not take into the reckoning.

On the face of this expanse of waters, and perfectly secure from fowlers, lie all day long, in the winter season, vast flocks of ducks, teals, and wigeons, of various denominations; where they preen and solace, and rest themselves, till towards sun-set, when they issue forth in little parties (for in their natural state they are all birds of the night) to feed in the brooks and meadows; returning again with the dawn of the morning. Had this lake an arm or two more, and were it planted round with thick covert (for now it is perfectly naked), it might make a valuable decoy.*

Yet neither its extent, nor the clearness of its water, nor the resort of various and curious fowls, nor its picturesque groups of cattle, can render this meer so remarkable as the great quantity of coins that were found in its bed about forty years ago. But, as such discoveries more properly belong to the antiquities of this place, I shall suppress all particulars for the present, till I enter professedly on my series of letters respecting the more remote history of this village and district.

LETTER 9

By way of supplement, I shall trouble you once more on this subject, to inform you that Wolmer, with her sister forest Ayles Holt, alias Alice Holt,[1] as it is called in old records, is held by grant from the crown for a term of years.

The grantees that the author remembers are Brigadier-General Emanuel Scroope Howe, and his lady, Ruperta, who was a natural daughter of Prince Rupert by Margaret Hughs; a Mr Mordaunt, of the Peterborough family, who married a dowager Lady Pembroke; Henry Bilson Legge and lady; and now Lord Stawel, their son.

The lady of General Howe lived to an advanced age, long surviving her husband; and, at her death, left behind her many curious pieces of mechanism of her father's constructing, who was a distinguished mechanic and artist,[2] as well as warrior; and, among the rest, a very complicated clock, lately in possession of Mr Elmer, the celebrated game-painter at Farnham, in the county of Surrey.

Though these two forests are only parted by a narrow range of

---

[1] 'In *Rot. Inquisit. de statu forest. in Scaccar.* 36. Ed. 3. it is called Aisholt.'
    In the same, 'Tit. Woolmer and Aisholt Hantisc. Dominus Rex habet unam capellam in haia sua de Kingesle.' 'Haia, sepes, sepimentum, parcus: a Gall. haie and haye.' Spelman's Glossary.
[2] This prince was the inventor of mezzotinto.*

enclosures, yet no two soils can be more different: for the Holt consists of a strong loam, of a miry nature, carrying a good turf, and abounding with oaks that grow to be large timber; while Wolmer is nothing but a hungry, sandy, barren waste.

The former, being all in the parish of Binsted, is about two miles in extent from north to south, and near as much from east to west; and contains within it many woodlands and lawns, and the great lodge where the grantees reside; and a smaller lodge called Goosegreen; and is abutted on by the parishes of Kingsley, Frinsham, Farnham, and Bentley; all of which have right of common.

One thing is remarkable; that, though the Holt has been of old well stocked with fallow-deer, unrestrained by any pales or fences more than a common hedge, yet they were never seen within the limits of Wolmer; nor were the red deer of Wolmer ever known to haunt the thickets or glades of the Holt.

At present the deer of the Holt are much thinned and reduced by the night-hunters, who perpetually harass them in spite of the efforts of numerous keepers, and the severe penalties that have been put in force against them as often as they have been detected, and rendered liable to the lash of the law. Neither fines nor imprisonments can deter them: so impossible is it to extinguish the spirit of sporting, which seems to be inherent in human nature.

General Howe turned out some German wild boars and sows in his forests, to the great terror of the neighbourhood; and, at one time, a wild bull or buffalo: but the country rose upon them and destroyed them.

A very large fall of timber, consisting of about one thousand oaks, has been cut this spring (*viz.* 1784) in the Holt forest; one fifth of which, it is said, belongs to the grantee, Lord Stawel. He lays claim also to the lop and top:* but the poor of the parishes of Binsted and Frinsham, Bentley and Kingsley, assert that it belongs to them; and, assembling in a riotous manner, have actually taken it all away. One man, who keeps a team, has carried home, for his share, forty stacks of wood. Forty-five of these people his Lordship has served with actions. These trees, which were very sound, and in high perfection, were winter-cut, *viz.* in February and March, before the bark would run.* In old times the Holt was estimated to be eighteen miles, computed measure, from water-carriage, *viz.* from the town of Chertsey, on the Thames; but now it is not half that distance, since the Wey is made navigable up to the town of Godalming in the county of Surrey.

LETTER 10

August 4, 1767

It has been my misfortune never to have had any neighbours whose studies have led them towards the pursuit of natural knowledge: so that, for want of a companion to quicken my industry and sharpen my attention, I have made but slender progress in a kind of information to which I have been attached from my childhood.

As to swallows (*hirundines rusticae*) being found in a torpid state during the winter in the isle of Wight, or any part of this country, I never heard any such account worth attending to. But a clergyman, of an inquisitive turn, assures me, that, when he was a great boy, some workmen, in pulling down the battlements of a church tower early in the spring, found two or three swifts (*hirundines apodes*) among the rubbish, which were, at first appearance, dead; but, on being carried toward the fire, revived. He told me that, out of his great care to preserve them, he put them in a paper-bag, and hung them by the kitchen fire, where they were suffocated.

Another intelligent person has informed me that, while he was a schoolboy at Brighthelmstone, in Sussex, a great fragment of the chalk-cliff fell down one stormy winter on the beach; and that many people found swallows among the rubbish: but, on my questioning him whether he saw any of those birds himself; to my no small disappointment, he answered me in the negative; but that others assured him they did.

Young broods of swallows began to appear this year on July the eleventh, and young martins (*hirundines urbicae*) were then fledged in their nests. Both species will breed again once. For I see by my *fauna* of last year, that young broods came forth so late as September the eighteenth. Are not these late hatchings more in favour of hiding than migration? Nay, some young martins remained in their nests last year so late as September the twenty-ninth; and yet they totally disappeared with us by the fifth of October.*

How strange is it that the swift, which seems to live exactly the same life with the swallow and house-martin, should leave us before the middle of August invariably! while the latter stay often till the middle of October; and once I saw numbers of housemartins on the seventh of November. The martins and red-wing fieldfares were flying in sight together; an uncommon assemblage of summer and winter-birds!

A little yellow bird (it is either a species of the *alauda trivialis*, or rather perhaps of the *motacilla trochilus*)* still continues to make a sibilous shivering noise in the tops of tall woods. The *stoparola* of Ray (for which we have as yet no name in these parts) is called, in your *Zoology*, the fly-catcher.* There is one circumstance characteristic of this bird, which seems to have escaped observation, and that is, it takes its stand on the top of some stake or post, from whence it springs forth on its prey, catching a fly in the air, and hardly ever touching the ground, but returning still to the same stand for many times together.

I perceive there are more than one species of the *motacilla trochilus*: Mr Derham supposes, in Ray's *Philos. Letters*, that he has discovered three.* In these there is again an instance of some very common birds that have as yet no English name.

Mr Stillingfleet makes a question whether the black-cap (*motacilla atricapilla*) be a bird of passage or not:* I think there is no doubt of it: for, in April, in the first fine weather, they come trooping, all at once, into these parts, but are never seen in the winter. They are delicate songsters.

Numbers of snipes breed every summer in some moory ground on the verge of this parish. It is very amusing to see the cock bird on wing at that time, and to hear his piping and humming notes.

I have had no opportunity yet of procuring any of those mice which I mentioned to you in town.* The person that brought me the last says they are plenty in harvest, at which time I will take care to get more; and will endeavour to put the matter out of doubt, whether it be a non-descript* species or not.

I suspect much there may be two species of water-rats. Ray says, and Linnaeus after him, that the water-rat is web-footed behind. Now I have discovered a rat on the banks of our little stream that is not web-footed, and yet is an exellent swimmer and diver: it answers exactly to the *mus amphibius* of Linnaeus (See *Syst. Nat.*) which he says 'natat in fossis & urinatur [swims in ditches, and dives]'. I should be glad to procure one 'plantis palmatis [with webbed feet]'. Linnaeus seems to be in a puzzle about his *mus amphibius*, and to doubt whether it differs from his *mus terrestris*; which if it be, as he allows, the '*mus agrestis capite grandi brachyuros*' of Ray, is widely different from the water-rat, both in size, make, and manner of life.

As to the *falco*, which I mentioned in town, I shall take the liberty to send it down to you into Wales;* presuming on your candour, that you will excuse me if it should appear as familiar to you as it is strange to me. Though mutilated 'qualem dices . . . antehac fuisse, tales cum sint reliquiae [you will tell from the remains what it was like originally]'!

It haunted a marshy piece of ground in quest of wild-ducks and snipes: but, when it was shot, had just knocked down a rook, which it was tearing in pieces. I cannot make it answer to any of our English hawks; neither could I find any like it at the curious exhibition of stuffed birds in Spring-Gardens.* I found it nailed up at the end of a barn, which is the countryman's museum.

The parish I live in is a very abrupt, uneven country, full of hills and woods, and therefore full of birds.

LETTER 11

Selborne, September 9, 1767

It will not be without impatience that I shall wait for your thoughts with regard to the *falco*; as to its weight, breadth, &c. I wish I had set them down at the time: but, to the best of my remembrance, it weighed two pounds and eight ounces, and measured, from wing to wing, thirty-eight inches. Its cere* and feet were yellow, and the circle of its eyelids a bright yellow. As it had been killed some days, and the eyes were sunk, I could make no good observation on the colour of the pupils and the irides.

The most unusual birds I ever observed in these parts were a pair of hoopoes (*upupa*), which came several years ago in the summer, and frequented an ornamented piece of ground, which joins to my garden, for some weeks. They used to march about in a stately manner, feeding in the walks, many times in the day; and seemed disposed to breed in my outlet;* but were frighted and persecuted by idle boys, who would never let them be at rest.

Three gross-beaks (*loxia coccothraustes*) appeared some years ago in my fields, in the winter; one of which I shot: since that, now and then one is occasionally seen in the same dead season.

A cross-bill (*loxia curvirostra*) was killed last year in this neighbourhood.

Our streams, which are small, and rise only at the end of the village, yield nothing but the bull's-head or miller's-thumb (*gobius fluviatilis capitatus*), the trout (*trutta fluviatilis*), the eel (*anguilla*), the lampern (*lampaetra parva et fluviatilis*), and the stickle-back (*pisciculus aculeatus*).

We are twenty miles from the sea, and almost as many from a great river, and therefore see but little of sea-birds. As to wild fowls, we have a few teams of ducks bred in the moors where the snipes breed; and multitudes of widgeons and teals in hard weather frequent our lakes in the forest.

Having some acquaintance with a tame brown owl, I find that it casts up the fur of mice, and the feathers of birds in pellets, after the manner of hawks: when full, like a dog, it hides what it cannot eat.

The young of the barn-owl are not easily raised, as they want a constant supply of fresh mice: whereas the young of the brown owl will eat indiscriminately all that is brought; snails, rats, kittens, puppies, magpies, and any kind of carrion or offal.

The house-martins have eggs still, and squab-young.* The last swift I observed was about the twenty-first of August; it was a straggler.

Red-starts, fly-catchers, white-throats, and *reguli non cristati*, still appear; but I have seen no black-caps lately.

I forgot to mention that I once saw, in Christ Church college quadrangle in Oxford, on a very sunny warm morning, a house martin flying about, and settling on the parapets, so late as the twentieth of November.

At present I know only two species of bats, the common *vespertilio murinus* and the *vespertilio auribus*.

I was much entertained last summer with a tame bat, which would take flies out of a person's hand. If you gave it any thing to eat, it brought its wings round before the mouth, hovering and hiding its head in the manner of birds of prey when they feed. The adroitness it shewed in shearing off the wings of the flies, which were always rejected, was worthy of observation, and pleased me much.

LONG EARED BAT.

Insects seemed to be most acceptable, though it did not refuse raw flesh when offered: so that the notion, that bats go down chimneys and gnaw men's bacon,* seems no improbable story. While I amused myself with this wonderful quadruped, I saw it several times confute the vulgar opinion, that bats when down on a flat surface cannot get on the wing again, by rising with great ease from the floor. It ran, I observed, with more dispatch than I was aware of; but in a most ridiculous and grotesque manner.

Bats drink on the wing, like swallows, by sipping the surface, as they play over pools and streams. They love to frequent waters, not only for the sake of drinking, but on account of insects, which are found over them in the greatest plenty. As I was going, some years ago, pretty late, in a boat from Richmond to Sunbury, on a warm summer's evening, I think I saw myriads of bats between the two places: the air swarmed with them all along the Thames, so that hundreds were in sight at a time.

I am, &c.

LETTER 12

November 4, 1767

Sir,

It gave me no small satisfaction to hear that the *falco*[1] turned out an uncommon one. I must confess I should have been better pleased to have heard that I had sent you a bird that you had never seen before; but that, I find, would be a difficult task.

I have procured some of the mice mentioned in my former letters, a young one and a female with young, both of which I have preserved in brandy. From the colour, shape, size, and manner of nesting, I make no doubt but that the species is non-descript.* They are much smaller, and more slender, than the *mus domesticus medius* of Ray; and have more of the squirrel or dormouse colour: their belly is white; a straight line along their sides divides the shades of their back and belly. They never enter into houses; are carried into ricks and barns with the sheaves; abound in harvest; and build their nests amidst the straws of the corn above the ground, and sometimes in thistles. They breed as many as eight at a litter, in a little round nest composed of the blades of grass or wheat.

One of these nests I procured this autumn, most artificially plat-ted, and composed of the blades of wheat; perfectly round, and about the size of a cricket-ball; with the aperture so ingeniously closed, that there was no discovering to what part it belonged. It was so compact and well filled, that it would roll across the table without being dis-composed, though it contained eight little mice that were naked and blind. As this nest was perfectly full, how could the dam come at her litter respectively so as to administer a teat to each? perhaps she opens different places for that purpose, adjusting them again when the business is over: but she could not possibly be contained herself in the ball with her young, which moreover would be daily increasing in bulk. This wonderful procreant cradle, an elegant instance of the efforts of instinct, was found in a wheat-field suspended in the head of a thistle.

A gentleman, curious in birds, wrote me word that his servant had shot one last January, in that severe weather, which he believed would

---

[1] This hawk proved to be the *falco peregrinus*; a variety.

puzzle me. I called to see it this summer, not knowing what to expect: but, the moment I took it in hand, I pronounced it the male *garrulus bohemicus* or German silk-tail, from the five peculiar crimson tags or points which it carries at the ends of five of the short remiges.* It cannot, I suppose, with any propriety, be called an English bird: and yet I see, by Ray's *Philosoph. Letters*, that great flocks of them, feeding on haws, appeared in this kingdom in the winter of 1685.*

The mention of haws puts me in mind that there is a total failure of that wild fruit, so conducive to the support of many of the winged nation. For the same severe weather, late in the spring, which cut off all the produce of the more tender and curious trees, destroyed also that of the more hardy and common.

Some birds, haunting with the missel-thrushes, and feeding on the berries of the yew-tree, which answered to the description of the *merula torquata* or ring-ouzel, were lately seen in this neighbourhood. I employed some people to procure me a specimen, but without success. See Letter 8.

Query—Might not Canary birds be naturalized to this climate, provided their eggs were put, in the spring, into the nests of some of their congeners,* as goldfinches, greenfinches, &c.? Before winter perhaps they might be hardened, and able to shift for themselves.

About ten years ago I used to spend some weeks yearly at Sunbury,* which is one of those pleasant villages lying on the Thames, near Hampton-court. In the autumn, I could not help being much amused with those myriads of the swallow kind which assemble in those parts. But what struck me most was, that, from the time they began to congregate, forsaking the chimnies and houses, they roosted every night in the osier-beds of the aits* of that river. Now this resorting towards that element, at that season of the year, seems to give some countenance to the northern opinion (strange as it is) of their retiring under water. A Swedish naturalist is so much persuaded of that fact, that he talks, in his calendar of Flora, as familiarly of the swallow's going under water in the beginning of September,* as he would of his poultry going to roost a little before sunset.

An observing gentleman in London writes me word that he saw an house-martin, on the twenty-third of last October, flying in and out of its nest in the Borough.* And I myself, on the twenty-ninth of last October (as I was travelling through Oxford), saw four or five swallows hovering round and settling on the roof of the county-hospital.

Now is it likely that these poor little birds (which perhaps had not been hatched but a few weeks) should, at that late season of the year, and from so midland a county, attempt a voyage to Goree or Senegal, almost as far as the equator?[1]*

I acquiesce entirely in your opinion—that, though most of the swallow kind may migrate, yet that some do stay behind and hide with us during the winter.

As to the short-winged soft-billed birds, which come trooping in such numbers in the spring, I am at a loss even what to suspect about them. I watched them narrowly this year, and saw them abound till about Michaelmas, when they appeared no longer. Subsist they cannot openly among us, and yet elude the eyes of the inquisitive: and, as to their hiding, no man pretends to have found any of them in a torpid state in the winter. But with regard to their migration, what difficulties attend that supposition! that such feeble bad fliers (who the summer long never flit but from hedge to hedge) should be able to traverse vast seas and continents in order to enjoy milder seasons amidst the regions of Africa!

LETTER 13

Selborne, Jan. 22, 1768

Sir,

As in one of your former letters you expressed the more satisfaction from my correspondence on account of my living in the most southerly county; so now I may return the compliment, and expect to have my curiosity gratified by your living much more to the North.

For many years past I have observed that towards Christmas vast flocks of chaffinches have appeared in the fields; many more, I used to think, than could be hatched in any one neighbourhood. But, when I came to observe them more narrowly, I was amazed to find that they seemed to me to be almost all hens. I communicated my suspicions to some intelligent neighbours, who, after taking pains about the matter, declared that they also thought them all mostly females; at least fifty to one. This extraordinary occurrence brought to my mind the

---

[1] See Adanson's *Voyage to Senegal*.

remark of Linnaeus; that 'before winter all their hen chaffinches migrate through Holland into Italy.' Now I want to know, from some curious person in the north, whether there are any large flocks of these finches with them in the winter, and of which sex they mostly consist? For, from such intelligence, one might be able to judge whether our female flocks migrate from the other end of the island, or whether they come over to us from the continent.

We have, in the winter, vast flocks of the common linnets; more, I think, than can be bred in any one district. These, I observe, when the spring advances, assemble on some tree in the sunshine, and join all in a gentle sort of chirping, as if they were about to break up their winter quarters and betake themselves to their proper summer homes. It is well known, at least, that the swallows and the fieldfares do congregate with a gentle twittering before they make their respective departure.

You may depend on it that the bunting, *emberiza miliaria*, does not leave this county in the winter. In January 1767 I saw several dozen of them, in the midst of a severe frost, among the bushes on the downs near Andover: in our woodland enclosed district it is a rare bird.

Wagtails, both white and yellow, are with us all the winter. Quails crowd to our southern coast, and are often killed in numbers by people that go on purpose.

Mr Stillingfleet, in his *Tracts*, says that 'if the wheatear (*aenanthe*) does not quit England, it certainly shifts places; for about harvest they are not to be found, where there was before great plenty of them.'* This well accounts for the vast quantities that are caught about that time on the south downs near Lewes, where they are esteemed a delicacy. There have been shepherds, I have been credibly informed, that have made many pounds in a season by catching them in traps. And though such multitudes are taken, I never saw (and I am well acquainted with those parts)* above two or three at a time: for they are never gregarious. They may perhaps migrate in general; and, for that purpose, draw towards the coast of Sussex in autumn: but that they do not all withdraw I am sure; because I see a few stragglers in many counties, at all times of the year, especially about warrens and stone quarries.

I have no acquaintance, at present, among the gentlemen of the navy: but have written to a friend, who was a sea-chaplain in the late war,* desiring him to look into his minutes, with respect to birds that settled on their rigging during their voyage up or down the channel. What Hasselquist says on that subject is remarkable: there were little short-winged birds frequently coming on board his ship all the way from our channel quite up to the Levant,* especially before squally weather.

What you suggest, with regard to Spain, is highly probable. The winters of Andalusia are so mild, that, in all likelihood, the soft-billed birds that leave us at that season may find insects sufficient to support them there.*

Some young man, possessed of fortune, health, and leisure, should make an autumnal voyage into that kingdom; and should spend a year there, investigating the natural history of that vast country. Mr Willughby[1] passed through that kingdom on such an errand; but he seems to have skirted along in a superficial manner and an ill humour, being much disgusted at the rude dissolute manners of the people.

I have no friend left now at Sunbury* to apply to about the swallows roosting on the aits of the Thames: nor can I hear any more about those birds which I suspected were *merulae torquatae*.

As to the small mice, I have farther to remark, that though they hang their nests for breeding up amidst the straws of the standing corn, above the ground; yet I find that, in the winter, they burrow deep in the earth, and make warm beds of grass: but their grand rendezvous seems to be in corn-ricks, into which they are carried at harvest. A neighbour housed an oat-rick lately, under the thatch of which were assembled near an hundred, most of which were taken; and some I saw. I measured them; and found that, from nose to tail, they were just two inches and a quarter, and their tails just two inches long. Two of them, in a scale, weighed down just one copper halfpenny, which is about the third of an ounce avoirdupois:* so that I suppose they are the smallest quadrupeds in this island. A full-grown *mus medius domesticus* weighs, I find, one ounce lumping weight,* which is more than six times as much as the mouse above; and measures from nose to rump four inches and a quarter, and the same in its tail. We have had a very severe frost and deep snow this month. My thermometer was one day fourteen degrees and an half below the freezing point, within doors. The tender evergreens were

---

[1] See Ray's *Travels*, p. 466.

injured pretty much. It was very providential that the air was still, and the ground well covered with snow, else vegetation in general must have suffered prodigiously. There is reason to believe that some days were more severe than any since the year 1739–40.*

I am, &c. &c.

LETTER 14

Selborne, March 12, 1768

Dear Sir,

If some curious gentleman would procure the head of a fallow-deer, and have it dissected, he would find it furnished with two spiracula, or breathing-places, besides the nostrils; probably analogous to the *puncta lachrymalia* [tear ducts] in the human head. When deer are thirsty they plunge their noses, like some horses, very deep under water, while in the act of drinking, and continue them in that situation for a considerable time: but, to obviate any inconveniency, they can open two vents, one at the inner corner of each eye, having a communication with the nose.* Here seems to be an extraordinary provision of nature worthy our attention; and which has not, that I know of, been noticed by any naturalist. For it looks as if these creatures would not be suffocated, though both their mouths and nostrils were stopped. This curious formation of the head may be of singular service to beasts of chase, by affording them free respiration: and no doubt these additional nostrils are thrown open when they are hard run.[1] Mr Ray observed that, at Malta, the owners slit up the nostrils of such asses as were hard worked: for they, being naturally strait or small, did not admit air sufficient to serve them when they travelled, or laboured, in that hot climate. And we know that grooms, and gentlemen of the turf, think large nostrils necessary, and a perfection, in hunters and running horses.

---

[1] In answer to this account, Mr Pennant sent me the following curious and pertinent reply. 'I was much surprised to find in the antelope something analogous to what you mention as so remarkable in deer. This animal also has a long slit beneath each eye, which can be opened and shut at pleasure. On holding an orange to one, the creature made as much use of those orifices as of his nostrils, applying them to the fruit, and seeming to smell it through them.'

Oppian, the Greek poet, by the following line, seems to have had some notion that stags have four spiracula:

Τετράδυμοι ῥῖνες, πίσυρες πνοιῇσι δίαυλοι.

Quadrifidae nares, quadruplices ad respirationem canales.

[Four nostrils, four passages for breathing.]

Opp., *Cyn[ogetica]* Lib. ii.1.181.

Writers, copying from one another, make Aristotle say that goats breathe at their ears; whereas he asserts just the contrary:—'Ἀλκμαίων γὰρ οὐκ ἀληθῆ λέγει, φάμενος ἀναπνεῖν τὰς αἰγὰς κατὰ τὰ ὦτά.' 'Alcmaeon does not advance what is true, when he avers that goats breathe through their ears.'—*History of Animals*, Book I. chap. xi.

LETTER 15

Selborne, March 30, 1768

Dear Sir,

Some intelligent country people have a notion that we have, in these parts, a species of the genus *mustelinum*, besides the weasel, stoat, ferret, and polecat; a little reddish beast, not much bigger than a field mouse, but much longer, which they call a cane.* This piece of intelligence can be little depended on; but farther inquiry may be made.

A gentleman in this neighbourhood had two milkwhite rooks in one nest. A booby of a carter, finding them before they were able to fly, threw them down and destroyed them, to the regret of the owner, who would have been glad to have preserved such a curiosity in his rookery. I saw the birds myself nailed against the end of a barn, and was surprised to find that their bills, legs, feet, and claws were milkwhite.

A shepherd saw, as he thought, some white larks on a down above my house this winter: were not these the *emberiza nivalis*, the snowflake of the *Brit. Zool.*?* No doubt they were.

A few years ago I saw a cock bullfinch in a cage, which had been caught in the fields after it was come to its full colours. In about a year it began to look dingy; and, blackening every succeeding year, it became coal-black at the end of four. Its chief food was hempseed. Such influence has food on the colour of animals! The pied and

WEESEL.

mottled colours of domesticated animals are supposed to be owing to high, various, and unusual food.

I had remarked, for years, that the root of the cuckoo-pint (*arum*) was frequently scratched out of the dry banks of hedges, and eaten in severe snowy weather. After observing, with some exactness, myself, and getting others to do the same, we found it was the thrush kind that searched it out. The root of the *arum* is remarkably warm and pungent.

Our flocks of female chaffinches have not yet forsaken us. The blackbirds and thrushes are very much thinned down by that fierce weather in January.

In the middle of February I discovered, in my tall hedges, a little bird that raised my curiosity: it was of that yellow-green colour that belongs to the *salicaria* kind, and, I think, was soft-billed.* It was no *parus*; and was too long and too big for the golden-crowned wren, appearing most like the largest willow-wren. It hung sometimes with its back downwards, but never continuing one moment in the same place. I shot at it, but it was so desultory that I misled my aim.*

I wonder that the stone curlew, *charadrius oedicnemus*, should be mentioned by the writers as a rare bird: it abounds in all the campaign* parts of Hampshire and Sussex, and breeds, I think, all the summer, having young ones, I know, very late in the autumn. Already they begin clamouring in the evening. They cannot, I think, with any propriety, be called, as they are by Mr Ray, '*circa aquas versantes* [circling over the water]'; for with us, by day at least, they haunt only the most dry, open, upland fields and sheep walks, far removed from water: what they may do in the night I cannot say. Worms are their usual food, but they also eat toads and frogs.

I can shew you some good specimens of my new mice. Linnaeus perhaps would call the species *mus minimus*.

LETTER 16

Selborne, April 18, 1768

Dear Sir,

The history of the stone curlew, *charadrius oedicnemus*, is as follows. It lays its eggs, usually two, never more than three, on the bare ground, without any nest, in the field; so that the countryman, in stirring his fallows,* often destroys them. The young run immediately from the egg like partridges, &c. and are withdrawn to some flinty field by the dam, where they sculk among the stones, which are their best security; for their feathers are so exactly of the colour of our grey spotted flints, that the most exact observer, unless he catches the eye of the young bird, may be eluded. The eggs are short and round; of a dirty white, spotted with dark bloody blotches. Though I might not be able, just when I pleased, to procure you a bird, yet I could shew you them almost any day; and any evening you may hear them round the village, for they make a clamour which may be heard a mile. *Oedicnemus* is a most apt and expressive name for them, since their legs seem swoln like those of a gouty man. After harvest I have shot them before the pointers* in turnip-fields.

I make no doubt but there are three species of the willow-wrens: two I know perfectly; but have not been able yet to procure the third. No two birds can differ more in their notes, and that constantly, than those two that I am acquainted with; for the one has a joyous, easy, laughing note; the other a harsh loud chirp. The former is every way larger, and three quarters of an inch longer, and weighs two drams* and an half; while the latter weighs but two: so the songster is one fifth heavier than the chirper. The chirper (being the first summer-bird of passage that is heard, the wryneck sometimes excepted) begins his two notes in the middle of March, and continues them through the spring and summer till the end of August, as appears by my journals. The legs of the larger of these two are flesh-coloured; of the less, black.

The grasshopper-lark began his sibilous note in my fields last Saturday. Nothing can be more amusing than the whisper of this little bird, which seems to be close by though at an hundred yards distance; and, when close at your ear, is scarce any louder than when a great way off. Had I not been a little acquainted with insects, and known that the grasshopper kind is not yet hatched, I should have hardly believed but that it had been a *locusta* whispering in the bushes. The country people laugh when you tell them that it is the note of a bird. It is a most artful creature, sculking in the thickest part of a bush; and will sing at a yard distance, provided it be concealed. I was obliged to get a person to go on the other side of the hedge where it haunted;* and then it would run, creeping like a mouse, before us for an hundred yards together, through the bottom of the thorns; yet it would not come into fair sight: but in a morning early, and when undisturbed, it sings on the top of a twig, gaping and shivering with its wings. Mr Ray himself had no knowledge of this bird, but received his account from Mr Johnson,* who apparently confounds it with the *reguli non cristati*, from which it is very distinct. See Ray's *Philos. Letters*, p. 108.

The fly-catcher (*stoparola*) has not yet appeared: it usually breeds in my vine. The redstart begins to sing: its note is short and imperfect, but is continued till about the middle of June. The willow-wrens (the smaller sort) are horrid pests in a garden, destroying the pease, cherries, currants, &c; and are so tame that a gun will not scare them.

A LIST of the SUMMER BIRDS of PASSAGE discovered in this neighbourhood, ranged somewhat in the order in which they appear:

	LINNAEI NOMINA*
Smallest willow-wren	*Motacilla trochilus*
Wryneck	*Jynx torquilla*
House-swallow	*Hirundo rustica*
Martin	*Hirundo urbica*
Sand-martin	*Hirundo riparia*
Cuckoo	*Cuculus canorus*
Nightingale	*Motacilla luscinia*
Blackcap	*Motacilla atricapilla*

Whitethroat	*Motacilla sylvia*
Middle willow-wren	*Motacilla trochilus*
Swift	*Hirundo apus*
Stone curlew?	*Charadrius oedicnemus?*
Turtle-dove?	*Turtur aldrovandi?*
Grasshopper-lark	*Alauda trivialis*
Landrail	*Rallus crex*
Largest willow-wren	*Motacilla trochilus*
Redstart	*Motacilla phoenicurus*
Goatsucker, or fern-owl	*Caprimulgus europaeus*
Fly-catcher	*Muscicapa grisola*

My countrymen talk much of a bird that makes a clatter with its bill against a dead bough, or some old pales, calling it a jarbird. I procured one to be shot in the very fact; it proved to be the *sitta europaea* (the nuthatch.) Mr Ray says that the less spotted wood-pecker does the same. This noise may be heard a furlong or more.

Now is the only time to ascertain the short-winged summer birds; for, when the leaf is out, there is no making any remarks on such a restless tribe; and, when once the young begin to appear, it is all confusion: there is no distinction of genus, species, or sex.

In breeding-time snipes play over the moors, piping and humming: they always hum as they are descending. Is not their hum ventriloquous* like that of the turkey? Some suspect it is made by their wings.

This morning I saw the golden-crowned wren, whose crown glitters like burnished gold. It often hangs like a titmouse, with its back downwards.

Yours, &c. &c.

LETTER 17

Selborne, June 18, 1768

Dear Sir,

On Wednesday last arrived your agreeable letter of June the 10th. It gives me great satisfaction to find that you pursue these studies still with such vigour, and are in such forwardness with regard to reptiles and fishes.*

The reptiles, few as they are, I am not acquainted with, so well as I could wish, with regard to their natural history. There is a degree of dubiousness and obscurity attending the propagation of this class of animals, something analagous to that of the cryptogamia in the sexual system of plants:* and the case is the same with regard to some of the fishes; as the eel, &c.

The method in which toads procreate and bring forth seems to be very much in the dark. Some authors say that they are viviparous: and yet Ray classes them among his oviparous animals; and is silent with regard to the manner of their bringing forth. Perhaps they may be ἔσω μὲν ᾠοτόκοι, ἔξω δε ζωοτόκοι [internally oviparous, but externally viviparous], as is known to be the case with the viper.

The copulation of frogs (or at least the appearance of it; for Swammerdam proves that the male has no *penis intrans*) is notorious to every body: because we see them sticking upon each other's backs for a month together in the spring: and yet I never saw, or read, of toads being observed in the same situation. It is strange that the matter with regard to the venom of toads has not been yet settled. That they are not noxious to some animals is plain: for ducks, buzzards, owls, stone curlews, and snakes, eat them, to my knowledge, with impunity. And I well remember the time, but was not eye-witness to the fact (though numbers of persons were) when a quack, at this village, ate a toad to make the country-people stare; afterwards he drank oil.*

I have been informed also, from undoubted authority, that some ladies (ladies you will say of peculiar taste) took a fancy to a toad, which they nourished summer after summer, for many years, till he grew to a monstrous size, with the maggots which turn to flesh flies. The reptile used to come forth every evening from an hole under the garden-steps; and was taken up, after supper, on the table to be fed. But at last a tame raven, kenning him* as he put forth his head, gave him such a severe stroke with his horny beak as put out one eye. After this accident the creature languished for some time and died.

I need not remind a gentleman of your extensive reading of the excellent account there is from Mr Derham, in Ray's *Wisdom of God in the Creation* (p. 365), concerning the migration of frogs from their breeding ponds. In this account he at once subverts that foolish opinion of their dropping from the clouds in rain; shewing that it is from the grateful coolness and moisture of those showers that they are tempted to set out on their travels, which they defer till those fall. Frogs are as yet in their tadpole state; but, in a few weeks, our lanes, paths, fields, will swarm for a few days with myriads of those emigrants, no larger than my little finger nail. Swammerdam gives a most accurate account of the method and situation in which the male impregnates the spawn of the female.* How wonderful is the oeconomy of Providence* with regard to the limbs of so vile a reptile! While it is an aquatic it has a fish-like tail, and no legs: as soon as the legs sprout, the tail drops off as useless, and the animal betakes itself to the land!

Merret, I trust, is widely mistaken when he advances that the *rana arborea** is an English reptile; it abounds in Germany and Switzerland.

It is to be remembered that the *salamandra aquatica* of Ray (the water-newt or eft) will frequently bite at the angler's bait, and is often caught on his hook. I used to take it for granted that the *salamandra aquatica* was hatched, lived, and died, in the water. But John Ellis, Esq. F.R.S. (the coralline Ellis) asserts, in a letter to the Royal Society, dated June the 5th, 1766, in his account of the mud inguana,* an amphibious *bipes* from South Carolina, that the water-eft, or newt, is only the larva of the land-eft, as tadpoles are of frogs. Lest I should be suspected to misunderstand his meaning, I shall give it in his own words. Speaking of the opercula or coverings to the gills of the mud inguana, he proceeds to say that 'The form of these pennated coverings approach very near to what I have some time ago observed in the larva or aquatic state of our English *lacerta*, known by the name of eft, or newt;* which serve them for coverings to their gills, and for fins to swim with while in this state; and which they lose, as well as the fins of their tails, when they change their state and become land animals, as I have observed, by keeping them alive for some time myself.'

Linnaeus, in his *Systema Naturae*, hints at what Mr Ellis advances more than once.

Providence has been so indulgent to us as to allow of but one venomous reptile of the serpent kind in these kingdoms, and that is the viper. As you propose the good of mankind to be an object of your publications, you will not omit to mention common sallad-oil as a sovereign remedy against the bite of the viper.* As to the blind worm (*anguis fragilis*, so called because it snaps in sunder with a small blow), I have found, on examination, that it is perfectly innocuous. A neighbouring yeoman (to whom I am indebted for some good hints) killed and opened a female viper about the twenty-seventh of May: he found her filled with a chain of eleven eggs, about the size of those of a blackbird; but none of them were advanced so far towards a state of maturity as to contain any rudiments of young. Though they are oviparous, yet they are viviparous also, hatching their young within their bellies, and then bringing them forth. Whereas snakes* lay chains of eggs every summer in my melon beds, in spite of all that my people can do to prevent them; which eggs do not hatch till the spring following, as I have often experienced. Several intelligent folks assure me that they have seen the viper open her mouth and admit her helpless young down her throat on sudden surprises, just as the female opossum does her brood into the pouch under her belly, upon the like emergencies; and yet the London viper-catchers insist on it, to Mr Barrington, that no such thing ever happens. The serpent kind eat, I believe, but once in a year; or, rather, but only just at one season of the year. Country people talk much of a water-snake, but, I am pretty sure, without any reason; for the common snake (*coluber natrix*) delights much to sport in the water, perhaps with a view to procure frogs and other food.

I cannot well guess how you are to make out your twelve species of reptiles, unless it be by the various species, or rather varieties, of our *lacerti*, of which Ray enumerates five. I have not had opportunity of ascertaining these; but remember well to have seen, formerly, several beautiful green lacerti on the sunny sandbanks near Farnham, in Surrey; and Ray admits there are such in Ireland.

LETTER 18

Selborne, July 27, 1768

Dear Sir,

I received your obliging and communicative letter of June the 28th, while I was on a visit at a gentleman's house,* where I had neither books to turn to, nor leisure to sit down, to return you an answer to many queries, which I wanted to resolve in the best manner that I am able.

A person, by my order, has searched our brooks, but could find no such fish as the *gasterosteus pungitius*: he found the *gasterosteus* *aculeatus* in plenty. This morning, in a basket, I packed a little earthen pot full of wet moss, and in it some sticklebacks, male and female; the females big with spawn: some lamperns; some bulls heads; but I could procure no minnows. This basket will be in Fleet-street by eight this evening; so I hope Mazel* will have them fresh and fair to-morrow morning. I gave some directions, in a letter, to what particulars the engraver should be attentive.

Finding, while I was on a visit, that I was within a reasonable distance of Ambresbury, I sent a servant over to that town, and procured several living specimens of loaches, which he brought, safe and brisk, in a glass decanter. They were taken in the gullies that were cut for watering the meadows. From these fishes (which measured from two to four inches in length) I took the following description: 'The loach, in its general aspect, has a pellucid appearance: its back is mottled with irregular collections of small black dots, not reaching much below the *linea lateralis*, as are the back and tail fins: a black line runs from each eye down to the nose; its belly is of a silvery white; the upper jaw projects beyond the lower, and is surrounded with six feelers, three on each side: its pectoral fins are large, its ventral much smaller; the fin behind its anus small; its dorsal-fin large, containing eight spines; its tail, where it joins to the tail-fin, remarkably broad, without any taperness, so as to be characteristic of this genus: the tail-fin is broad, and square at the end. From the breadth and muscular strength of the tail it appears to be an active nimble fish.'

In my visit I was not very far from Hungerford, and did not forget to make some inquiries concerning the wonderful method of curing cancers by means of toads.* Several intelligent persons, both gentry and clergy, do, I find, give a great deal of credit to what was asserted in the papers: and I myself dined with a clergyman who seemed to be persuaded that what is related is matter of fact; but, when I came to attend to his account, I thought I discerned circumstances which did not a little invalidate the woman's story of the manner in which she came by her skill. She says of herself 'that, labouring under a virulent cancer, she went to some church where there was a vast crowd: on going into a pew, she was accosted by a strange clergyman; who, after expressing compassion for her situation, told her that if she would make such an application of living toads as is mentioned she would be well.' Now is it likely that this unknown gentleman should express so much tenderness for this single sufferer, and not feel any for the many thousands that daily languish under this terrible disorder? Would he not have made use of this invaluable nostrum for his own emolument;* or, at least, by some means of publication or other, have found a method of making it public for the good of mankind? In short, this woman (as it appears to me) having set up for a cancer-doctress, finds it expedient to amuse the country with this dark and mysterious relation.

The water-eft has not, that I can discern, the least appearance of any gills; for want of which it is continually rising to the surface of the water to take in fresh air. I opened a big-bellied one indeed, and found it full of spawn. Not that this circumstance at all invalidates the assertion that they are larvae: for the larvae of insects are full of eggs, which they exclude* the instant they enter their last state. The water-eft is continually climbing over the brims of the vessel, within which we keep it in water, and wandering away: and people every summer see numbers crawling, out of the pools where they are hatched, up the dry banks. There are varieties of them, differing in colour; and some have fins up their tail and back, and some have not.

LETTER 19

Selborne, Aug. 17, 1768

Dear Sir,

I have now, past dispute, made out three distinct species of the willow-wrens (*motacillae trochili*) which constantly and invariably use distinct notes.* But, at the same time, I am obliged to confess that I know nothing of your willow-lark.[1] In my letter of April the 18th, I had told you peremptorily* that I knew your willow-lark, but had not seen it then: but, when I came to procure it, it proved, in all respects, a very *motacilla trochilus*; only that it is a size larger than the two other, and the yellow-green of the whole upper part of the body is more vivid, and the belly of a clearer white. I have specimens of the three sorts now lying before me; and can discern that there are three gradations of sizes, and that the least has black legs, and the other two flesh-coloured ones. The yellowest bird is considerably the largest,

[1] *Brit. Zool.* edit. 1776, octavo, p. 381.

and has its quill-feathers and secondary feathers tipped with white, which the others have not. This last haunts only the tops of trees in high beechen woods, and makes a sibilous grasshopper-like noise, now and then, at short intervals, shivering a little with its wings when it sings; and is, I make no doubt now, the *regulus non cristatus* of Ray; which he says 'cantat voce stridulâ locustae [sings with the rasping sound of a grasshopper. See *Ornithology*, 1678]'. Yet this great ornithologist never suspected that there were three species.

LETTER 20

Selborne, October 8, 1768

It is, I find, in zoology as it is in botany: all nature is so full, that that district produces the greatest variety which is the most examined. Several birds, which are said to belong to the north only, are, it seems, often in the south. I have discovered this summer three species of birds with us, which writers mention as only to be seen in the northern counties. The first that was brought me (on the 14th of May), was the sandpiper, *tringa hypoleucus*: it was a cock bird, and haunted the banks of some ponds near the village; and, as it had a companion, doubtless intended to have bred near that water. Besides, the owner has told me since, that, on recollection, he has seen some of the same birds round his ponds in former summers.

The next bird that I procured (on the 21st of May) was a male redbacked butcher bird,* *lanius collurio*. My neighbour, who shot it, says that it might easily have escaped his notice, had not the outcries and chattering of the white-throats and other small birds drawn his attention to the bush where it was: its craw was filled with the legs and wings of beetles.

The next rare birds (which were procured for me last week) were some ring-ousels, *turdi torquati*.

This week twelve months a gentleman from London,* being with us, was amusing himself with a gun, and found, he told us, on an old yew hedge where there were berries, some birds like blackbirds, with rings of white round their necks: a neighbouring farmer also at the same time observed the same; but, as no specimens were procured, little notice was taken. I mentioned this circumstance to you in my letter of November the 4th, 1767: (you however paid but small regard to what I said, as I had not seen these birds myself): but last week the aforesaid farmer, seeing a large flock, twenty or thirty of these birds, shot two cocks and two hens: and says, on recollection, that he remembers to have observed these birds again last spring, about Lady-day,* as it were, on their return to the north. Now perhaps these ousels are not the ousels of the north of England, but belong to the more northern parts of Europe; and may retire before the excessive rigor of the frosts in those parts; and return to breed in the spring, when the cold abates. If this be the case, here is discovered a new bird of winter passage, concerning whose migrations the writers are silent: but if these birds should prove the ousels of the north of England, then here is a migration disclosed within our own kingdom never before remarked. It does not yet appear whether they retire beyond the bounds of our island to the south; but it is most probable that they usually do, or else one cannot suppose that they would have continued so long unnoticed in the southern counties. The ousel is larger than a blackbird, and feeds on haws; but last autumn (when there were no haws) it fed on yew-berries: in the spring it feeds on ivy-berries, which ripen only at that season, in March and April.

I must not omit to tell you (as you have been so lately on the study of reptiles) that my people, every now and then of late, draw up with a bucket of water from my well, which is 63 feet deep, a large black warty lizard* with a fin-tail and yellow belly. How they first came down at that depth, and how they were ever to have got out thence without help, is more than I am able to say.

My thanks are due to you for your trouble and care in the examination of a buck's head. As far as your discoveries reach at present, they seem much to corroborate my suspicions; and I hope Mr —— may find reason to give his decision in my favour; and then, I think, we may advance this extraordinary provision of nature as a new instance of the wisdom of God in the creation.*

As yet I have not quite done with my history of the *oedicnemus*, or stone-curlew; for I shall desire a gentleman in Sussex* (near whose house these birds congregate in vast flocks in the autumn) to observe nicely when they leave him, (if they do leave him) and when they return again in the spring: I was with this gentleman lately, and saw several single birds.

LETTER 21

Selborne, Nov. 28, 1768

Dear Sir,

With regard to the *oedicnemus*, or stone-curlew, I intend to write very soon to my friend near Chichester,* in whose neighbourhood these birds seem most to abound; and shall urge him to take particular notice when they begin to congregate, and afterwards to watch them most narrowly whether they do not withdraw themselves during the dead of the winter. When I have obtained information with respect to this circumstance, I shall have finished my history of the stone-curlew; which I hope will prove to your satisfaction, as it will be, I trust, very near the truth. This gentleman, as he occupies a large farm of his own, and is abroad early and late, will be a very proper spy upon the motions of these birds: and besides, as I have prevailed on him to buy the *Naturalist's Journal** (with which he is much delighted), I shall expect that he will be very exact in his dates. It is very extraordinary, as you observe, that a bird so common with us should never straggle to you.*

And here will be the properest place to mention, while I think of it, an anecdote which the above-mentioned gentleman told me when I was last at his house; which was that, in a warren joining to his outlet,* many daws (*corvi monedulae*) build every year in the rabbit-burrows under ground. The way he and his brothers used to take their nests, while they were boys, was by listening at the mouths of the holes; and, if they heard the young ones cry, they twisted the nest out with a forked stick. Some water-fowls (*viz.* the puffins) breed, I know, in that manner; but I should never have suspected the daws of building in holes on the flat ground.

Another very unlikely spot is made use of by daws as a place to breed in, and that is Stonehenge. These birds deposit their nests in the interstices between the upright and the impost stones* of that amazing work of antiquity: which circumstance alone speaks the prodigious height of the upright stones, that they should be tall enough to secure those nests from the annoyance of shepherd-boys, who are always idling round that place.

One of my neighbours last Saturday, November the 26th, saw a martin in a sheltered bottom:* the sun shone warm, and the bird was hawking briskly after flies. I am now perfectly satisfied that they do not all leave this island in the winter.

You judge very right, I think, in speaking with reserve and caution concerning the cures done by toads:* for, let people advance what they will on such subjects, yet there is such a propensity in mankind towards deceiving and being deceived, that one cannot safely relate any thing from common report, especially in print, without expressing some degree of doubt and suspicion.

Your approbation, with regard to my new discovery of the migration of the ring-ousel, gives me satisfaction; and I find you concur with me in suspecting that they are foreign birds which visit us. You will be sure, I hope, not to omit to make inquiry whether your ring-ousels leave your rocks in the autumn. What puzzles me most, is the very short stay they make with us; for in about three weeks they are all gone. I shall be very curious to remark whether they will call on us at their return in the spring, as they did last year.

I want to be better informed with regard to icthyology.* If fortune had settled me near the sea-side, or near some great river, my natural propensity would soon have urged me to have made myself acquainted with their productions: but as I have lived mostly in inland parts, and in an upland district, my knowledge of fishes extends little farther than to those common sorts which our brooks and lakes produce.

I am, &c.

LETTER 22

Selborne, Jan. 2, 1769

Dear Sir,

As to the peculiarity of jackdaws building with us under the ground in rabbit-burrows, you have, in part, hit upon the reason; for, in reality, there are hardly any towers or steeples in all this country. And perhaps, Norfolk excepted, Hampshire and Sussex are as meanly furnished with churches as almost any counties in the kingdom. We have many livings* of two or three hundred pounds a year, whose houses of worship make little better appearance than dovecots. When I first saw Northamptonshire, Cambridgeshire and Huntingdonshire, and the fens of Lincolnshire, I was amazed at the number of spires which presented themselves in every point of view.* As an admirer of prospects,* I have reason to lament this want in my own country; for such objects are very necessary ingredients in an elegant landscape.

What you mention with respect to reclaimed toads raises my curiosity. An ancient author, though no naturalist, has well remarked that 'Every kind of beasts, and of birds, and of serpents, and things in the sea, is tamed, and hath been tamed, of mankind.'[1]

It is a satisfaction to me to find that a green lizard has actually been procured for you in Devonshire; because it corroborates my discovery, which I made many years ago, of the same sort, on a sunny sandbank near Farnham, in Surrey. I am well acquainted with the south hams* of Devonshire; and can suppose that district, from its southerly situation, to be a proper habitation for such animals in their best colours.

Since the ring-ousels of your vast mountains do certainly not forsake them against winter, our suspicions that those which visit this neighbourhood about Michaelmas are not English birds, but driven from the more northern parts of Europe by the frosts, are still more reasonable; and it will be worth your pains to endeavour to trace from whence they come, and to inquire why they make so very short a stay.

In your account of your error with regard to the two species of herons,* you incidentally gave me great entertainment in your description of the heronry at Cressi-hall; which is a curiosity I never could manage to see. Fourscore nests of such a bird on one tree is a rarity which I would ride half as many miles to have a sight of. Pray be sure to tell me in your next whose seat Cressi-hall is,* and near what town it lies.[2] I have often thought that those vast extents of fens have never been sufficiently explored. If half a dozen gentlemen, furnished with a good strength of water-spaniels, were to beat them over for a week, they would certainly find more species.

There is no bird, I believe, whose manners I have studied more than that of the *caprimulgus* (the goat-sucker), as it is a wonderful and curious creature: but I have always found that though sometimes it may chatter as it flies, as I know it does, yet in general it utters its jarring note sitting on a bough; and I have for many an half hour watched it as it sat with its under mandible quivering, and particularly this summer. It perches usually on a bare twig, with its head lower

---

[1] James 3: 7.
[2] Cressi-hall is near Spalding, in Lincolnshire.

than its tail, in an attitude well expressed by your draughtsman in the folio *British Zoology*.* This bird is most punctual in beginning its song exactly at the close of day; so exactly that I have known it strike up more than once or twice just at the report of the Portsmouth evening gun,* which we can hear when the weather is still. It appears to me past all doubt that its notes are formed by organic impulse, by the powers of the parts of its windpipe, formed for sound, just as cats pur. You will credit me, I hope, when I assure you that, as my neighbours were assembled in an hermitage[1] on the side of a steep hill where we drink tea, one of these churn-owls came and settled on the cross of that little straw edifice and began to chatter, and continued his note for many minutes: and we were all struck with wonder to find that the organs of that little animal, when put in motion, gave a sensible vibration to the whole building!* This bird also sometimes makes a small squeak, repeated four or five times; and I have observed that to happen when the cock has been pursuing the hen in a toying way through the boughs of a tree.

It would not be at all strange if your bat, which you have procured, should prove a new one, since five species have been found in a neighbouring kingdom. The great sort that I mentioned is certainly a nondescript:* I saw but one this summer, and that I had no opportunity of taking.

Your account of the Indian-grass was entertaining. I am no angler myself; but inquiring of those that are, what they supposed that part of their tackle to be made of? they replied 'of the intestines of a silkworm.'*

Though I must not pretend to great skill in entomology, yet I cannot say that I am ignorant of that kind of knowledge: I may now and then perhaps be able to furnish you with a little information.

The vast rains ceased with us much about the same time as with you, and since we have had delicate weather. Mr Barker, who has measured the rain for more than thirty years, says, in a late letter, that more has fallen this year than in any he ever attended to; though, from July 1763 to January 1764, more fell than in any seven months of this year.

---

[1] See the vignette in this book.*

LETTER 23

Selborne, February 28, 1769

Dear Sir,

It is not improbable that the Guernsey lizard and our green lizards may be specifically the same; all that I know is, that, when some years ago many Guernsey lizards were turned loose in Pembroke college garden, in the university of Oxford,* they lived a great while, and seemed to enjoy themselves very well, but never bred. Whether this circumstance will prove any thing either way I shall not pretend to say.

I return you thanks for your account of Cressi-hall; but recollect, not without regret, that in June 1746 I was visiting for a week together at Spalding, without ever being told that such a curiosity was just at hand. Pray send me word in your next what sort of tree it is that contains such a quantity of herons' nests;* and whether the heronry consists of a whole grove or wood, or only of a few trees.

It gave me satisfaction to find we accorded so well about the *caprimulgus*:* all I contended for was to prove that it often chatters sitting as well as flying; and therefore the noise was voluntary, and from organic impulse, and not from the resistance of the air against the hollow of its mouth and throat.

If ever I saw any thing like actual migration, it was last Michaelmas-day. I was travelling, and out early in the morning: at first there was a vast fog; but, by the time that I was got seven or eight miles from home towards the coast, the sun broke out into a delicate warm day. We were then on a large heath or common, and I could discern, as the mist began to break away, great numbers of swallows (*hirundines rusticae*) clustering on the stunted shrubs and bushes, as if they had roosted there all night. As soon as the air became clear and pleasant they all were on the wing at once; and, by a placid and easy flight, proceeded on southward towards the sea: after this I did not see any more flocks, only now and then a straggler.*

I cannot agree with those persons that assert that the swallow kind disappear some and some gradually, as they come, for the bulk of them seem to withdraw at once: only some stragglers stay behind a long while, and do never, there is the greatest reason to believe, leave this island. Swallows seem to lay themselves up, and to come forth in a warm day, as bats do continually of a warm evening, after they have disappeared for weeks. For a very respectable gentleman assured me that, as he was walking with some friends under Merton-wall* on a remarkably hot noon, either in the last week in December or the first week in January, he espied three or four swallows huddled together on the moulding of one of the windows of that college. I have frequently remarked that swallows are seen later at Oxford than else-where: is it owing to the vast massy buildings of that place, to the many waters round it, or to what else?

When I used to rise in a morning last autumn, and see the swal-lows and martins clustering on the chimneys and thatch of the neigh-bouring cottages, I could not help being touched with a secret delight, mixed with some degree of mortification: with delight, to observe with how much ardour and punctuality those poor little birds obeyed the strong impulse towards migration, or hiding, imprinted on their minds by their great Creator; and with some degree of mortification, when I reflected that, after all our pains and inquiries, we are yet not quite certain to what regions they do migrate; and are still farther embarrassed to find that some do not actually migrate at all.

These reflections made so strong an impression on my imagin-ation, that they became productive of a composition that may perhaps amuse you for a quarter of an hour when next I have the honour of writing to you.

LETTER 24

Selborne, May 29, 1769

Dear Sir,

The *scarabaeus fullo* I know very well, having seen it in collections; but have never been able to discover one wild in its natural state. Mr Banks told me he thought it might be found on the seacoast.*

On the thirteenth of April I went to the sheep-down, where the ring-ousels have been observed to make their appearance at spring and fall, in their way perhaps to the north or south; and was much pleased to see three birds about the usual spot. We shot a cock and a hen; they were plump and in high condition. The hen had but very small rudiments of eggs within her, which proves they are late breeders; whereas those species of the thrush kind that remain with us the whole year have fledged young before that time. In their crops was nothing very distinguishable, but somewhat that seemed like blades of vegetables nearly digested. In autumn they feed on haws and yew-berries, and in the spring on ivy-berries. I dressed* one of these birds, and found it juicy and well flavoured. It is remarkable that they make but a few days stay in their spring visit, but rest near a fortnight at Michaelmas. These birds, from the observations of three springs and two autumns, are most punctual in their return; and exhibit a new migration unnoticed by the writers, who supposed they never were to be seen in any of the southern counties.*

One of my neighbours lately brought me a new *salicaria*, which at first I suspected might have proved your willow-lark,[1] but, on a nicer examination, it answered much better to the description of that species which you shot at Revesby, in Lincolnshire.* My bird I describe thus: 'It is a size less than the grasshopper-lark; the head, back, and coverts of the wings, of a dusky brown, without those dark spots of the grasshopper-lark; over each eye is a milkwhite stroke; the chin and throat are white, and the under parts of a yellowish white; the rump is tawny, and the feathers of the tail sharp-pointed; the bill is dusky and sharp, and the legs are dusky; the hinder claw long and crooked.' The person that shot it says that it sung so like a reed-sparrow that he took it for one; and that it sings all night: but this account merits farther inquiry. For my part, I suspect it is a second sort of *locustella*,* hinted at by Dr Derham in Ray's *Letters*: see p. 108. He also procured me a grasshopper-lark.

The question that you put with regard to those genera of animals that are peculiar to America, *viz*. how they came there, and whence? is too puzzling for me to answer; and yet so obvious as often to have struck me with wonder. If one looks into the writers on that subject little satisfaction is to be found. Ingenious men will readily advance plausible arguments to support whatever theory they shall chuse to maintain; but then the misfortune is, every one's hypothesis is each as good as another's, since they are all founded on conjecture. The late writers of this sort, in whom may be seen all the arguments of those

---

[1] For this *salicaria* see letter August 30, 1769.*

that have gone before, as I remember, stock America from the western coast of Africa and the south of Europe; and then break down the Isthmus that bridged over the Atlantic.* But this is making use of a violent piece of machinery: it is a difficulty worthy of the interposition of a god! 'Incredulus odi [I disbelieve and hate].'

TO THOMAS PENNANT, ESQUIRE

THE NATURALIST'S SUMMER-EVENING WALK

> . . . equidem credo, quia fit divinitus illis
> Ingenium.
>
> [I certainly believe they possess divine
> wisdom.]   [Virgil, *Georgics* I]

When day declining sheds a milder gleam,
What time the may-fly[1] haunts the pool or stream;
When the still owl skims round the grassy mead,
What time the timorous hare limps forth to feed;
Then be the time to steal adown the vale,
And listen to the vagrant[2] cuckoo's tale;
To hear the clamorous[3] curlew call his mate,
Or the soft quail his tender pain relate;
To see the swallow sweep the dark'ning plain
Belated, to support her infant train;
To mark the swift in rapid giddy ring
Dash round the steeple, unsubdu'd of wing:
Amusive* birds!—say where your hid retreat
When the frost rages and the tempests beat;
Whence your return, by such nice instinct led,
When spring, soft season, lifts her bloomy head?

---

[1] The angler's may-fly, the *ephemera vulgata* Linn. comes forth from its aurelia* state, and emerges out of the water about six in the evening, and dies about eleven at night, determining the date of its fly state in about five or six hours. They usually begin to appear about the 4th of June, and continue in succession for near a fortnight. See Swammerdam, Derham, Scopoli, &c.

[2] Vagrant cuckoo; so called because, being tied down by no incubation or attendance about the nutrition of its young, it wanders without control.

[3] *Charadrius oedicnemus.*

Such baffled searches mock man's prying pride,
The GOD of NATURE is your secret guide!
  While deep'ning shades obscure the face of day
To yonder bench leaf-shelter'd let us stray,
'Till blended objects fail the swimming sight,
And all the fading landscape sinks in night;
To hear the drowsy dor come brushing by
With buzzing wing, or the shrill[1] cricket cry;
To see the feeding bat glance through the wood;
To catch the distant falling of the flood;
While o'er the cliff th' awaken'd churn-owl hung
Through the still gloom protracts his chattering song;
While high in air, and pois'd upon his wings,
Unseen, the soft enamour'd[2] woodlark sings:
These, NATURE'S works, the curious mind employ,
Inspire a soothing melancholy joy:
As fancy warms, a pleasing kind of pain
Steals o'er the cheek, and thrills the creeping vein!
  Each rural sight, each sound, each smell, combine;
The tinkling sheep-bell, or the breath of kine;
The new-mown hay that scents the swelling breeze,
Or cottage-chimney smoking through the trees.
  The chilling night-dews fall:—away, retire;
For see, the glow-worm lights her amorous fire[3]!
Thus, e'er night's veil had half obscur'd the sky,
Th' impatient damsel hung her lamp on high:
True to the signal, by love's meteor led,
Leander hasten'd to his Hero's bed.[4]

I am, &c.

---

[1] *Gryllus campestris.*

[2] In hot summer nights woodlarks soar to a prodigious height and hang singing in the air.

[3] The light of the female glow-worm (as she often crawls up the stalk of a grass to make herself more conspicuous) is a signal to the male, which is a slender dusky *scarabaeus.*

[4] See the story of Hero and Leander.*

LETTER 25

Selborne, Aug. 30, 1769

Dear Sir,

It gives me satisfaction to find that my account of the ousel migration pleases you. You put a very shrewd question when you ask me how I know that their autumnal migration is southward? Was not candour and openness the very life of natural history, I should pass over this query just as a sly commentator does over a crabbed* passage in a classic; but common ingenuousness obliges me to confess, not without some degree of shame, that I only reasoned in that case from analogy. For as all other autumnal birds migrate from the northward to us, to partake of our milder winters, and return to the northward again when the rigorous cold abates, so I concluded that the ring-ousels did the same, as well as their congeners the fieldfares; and especially as ring-ousels are known to haunt cold mountainous countries: but I have good reason to suspect since that they may come to us from the westward; because I hear, from very good authority, that they breed on Dartmore; and that they forsake that wild district about the time that our visitors appear, and do not return till late in the spring.

I have taken a great deal of pains about your *salicaria* and mine, with a white stroke over its eye and a tawny rump. I have surveyed it alive and dead, and have procured several specimens; and am perfectly persuaded myself (and trust you will soon be convinced of the same) that it is no more nor less than the *passer arundinaceus minor* of Ray. This bird, by some means or other, seems to be entirely omitted in the *British Zoology*; and one reason probably was because it is so strangely classed in Ray, who ranges it among his *picis affines*. It ought no doubt to have gone among his *aviculae caudâ unicolore*, and among your slender-billed small birds of the same division. Linnaeus might with great propriety have put it into his genus of *motacilla*; and the *motacilla salicaria* of his *fauna suecica* seems to come the nearest to it. It is no uncommon bird, haunting the sides of ponds and rivers where there is covert, and the reeds and sedges of moors. The country people in some places call it the sedge-bird. It sings incessantly night and day during the breeding-time, imitating the note of a sparrow, a swallow, a sky-lark; and has a strange hurrying manner in its song.

My specimens correspond most minutely to the description of your fen *salicaria* shot near Revesby.* Mr Ray has given an excellent characteristic of it when he says, 'Rostrum & pedes in hâc aviculâ multò majores sunt quâm pro corporis ratione [The beak and feet of this bird are very much larger in proportion to its body].' See letter May 29, 1769.*

I have got you the egg of an *oedicnemus*, or stone-curlew, which was picked up in a fallow on the naked ground: there were two; but the finder inadvertently crushed one with his foot before he saw them.

When I wrote to you last year on reptiles, I wish I had not forgot to mention the faculty that snakes have of stinking *se defendendo* [in self-defence]. I knew a gentleman who kept a tame snake, which was in its person as sweet as any animal while in good humour and unalarmed; but as soon as a stranger, or a dog or cat, came in, it fell to hissing, and filled the room with such nauseous effluvia as rendered it hardly supportable. Thus the squnck, or stonck,* of Ray's *Synop. Quadr.* is an innocuous and sweet animal; but, when pressed hard by dogs and men, it can eject such a most pestilent and fetid smell and excrement, that nothing can be more horrible.

A gentleman sent me lately a fine specimen of the *lanius minor cinerascens cum maculâ in scapulis albâ, Raii* [lesser ashy shrike with white spots on its shoulders]; which is a bird that, at the time of your publishing your two first volumes of *British Zoology*, I find you had not seen. You have described it well from Edward's drawing.*

LETTER 26

Selborne, December 8, 1769

Dear Sir,

I was much gratified by your communicative letter on your return from Scotland, where you spent, I find, some considerable time, and gave yourself good room to examine the natural curiosities of that extensive kingdom, both those of the islands, as well as those of the highlands. The usual bane of such expeditions is hurry; because men seldom allot themselves half the time they should do: but, fixing on a day for their return, post from place to place, rather as if they were on a journey that required dispatch, than as philosophers investigating the works of nature. You must have made, no doubt, many discoveries, and laid up a good fund of materials for a future edition of the *British Zoology*; and will have no reason to repent that you have bestowed so much pains on a part of Great-Britain that perhaps was never so well examined before.

It has always been matter of wonder to me that fieldfares, which are so congenerous to thrushes and blackbirds, should never chuse to breed in England: but that they should not think even the highlands cold and northerly, and sequestered enough, is a circumstance still more strange and wonderful. The ring-ousel, you find, stays in Scotland the whole year round; so that we have reason to conclude that those migrators that visit us for a short space every autumn do not come from thence.

And here, I think, will be the proper place to mention that those birds were most punctual again in their migration this autumn, appearing, as before, about the 30th of September: but their flocks were larger than common, and their stay protracted somewhat beyond the usual time. If they came to spend the whole winter with us, as some of their congeners do, and then left us, as they do, in spring, I should not be so much struck with the occurrence, since it would be similar to that of the other winter birds of passage; but when I see them for a fortnight at Michaelmas, and again for about a week in the middle of April, I am seized with wonder, and long to be informed whence these travellers come, and whither they go, since they seem to use our hills merely as an inn or baiting place.*

Your account of the greater brambling, or snow-fleck,* is very amusing; and strange it is that such a short-winged bird should delight in such perilous voyages over the northern ocean! Some country people in the winter time have every now and then told me that they have seen two or three white larks on our downs; but, on considering the matter, I begin to suspect that these are some stragglers of the birds we are talking of, which sometimes perhaps may rove so far to the southward.

It pleases me to find that white hares are so frequent on the Scottish mountains, and especially as you inform me that it is a distinct species; for the quadrupeds of Britain are so few, that every new species is a great acquisition.

The eagle-owl, could it be proved to belong to us, is so majestic

WHITE HARE.   RABBET

WATER SHREW MOUSE.

a bird, that it would grace our fauna much. I never was informed before where wild-geese are known to breed.

You admit, I find, that I have proved your fen *salicaria* to be the lesser reed-sparrow of Ray: and I think you may be secure that I am right; for I took very particular pains to clear up that matter, and had some fair specimens; but, as they were not well preserved, they are decayed already. You will, no doubt, insert it in its proper place in your next edition. Your additional plates will much improve your work.*

De Buffon, I know, has described the water shrew-mouse: but still I am pleased to find you have discovered it in Lincolnshire, for the reason I have given in the article of* the white hare.

As a neighbour was lately plowing in a dry chalky field, far removed from any water, he turned out a water-rat, that was curiously laid up in an *hybernaculum** artificially formed of grass and leaves. At one end of the burrow lay above a gallon* of potatoes regularly stowed, on which it was to have supported itself for the winter. But the difficulty with me is how this *amphibius mus* came to fix its winter station at such a distance from the water. Was it determined in its choice of that place by the mere accident of finding the potatoes which were planted there; or is it the constant practice of the aquatic-rat to forsake the neighbourhood of the water in the colder months?

EAGLE OWL

Though I delight very little in analogous reasoning, knowing how fallacious it is with respect to natural history; yet, in the following instance, I cannot help being inclined to think it may conduce towards the explanation of a difficulty that I have mentioned before, with respect to the invariable early retreat of the *hirundo apus*, or swift, so many weeks before its congeners; and that not only with us, but also in Andalusia, where they also begin to retire about the beginning of August.*

The great large bat[1] (which by the by is at present a non-descript in England,* and what I have never been able yet to procure) retires or migrates very early in the summer: it also ranges very high for its food, feeding in a different region of the air; and that is the reason I never could procure one. Now this is exactly the case with the swifts; for they take their food in a more exalted region than the other species, and are very seldom seen hawking for flies near the ground, or over the surface of the water. From hence I would conclude that these hirundines, and the larger bats, are supported by some sorts of high-flying gnats, scarabs, or *phalaenae*,* that are of short continuance; and that the short stay of these strangers is regulated by the defect* of their food.

By my journal it appears that curlews clamoured on to October the thirty-first; since which I have not seen or heard any. Swallows were observed on to November the third.

LETTER 27

Selborne, Feb. 22, 1770

Dear Sir,

Hedge-hogs abound in my gardens and fields. The manner in which they eat their roots of the plantain in my grass-walks is very curious: with their upper mandible, which is much longer than their lower, they bore under the plant, and so eat the root off upwards, leaving the tuft of leaves untouched. In this respect they are serviceable, as they destroy a very troublesome weed; but they deface the

---

[1] The little bat appears almost every month in the year; but I have never seen the large ones till the end of April, nor after July. They are most common in June, but never in any plenty: are a rare species with us.

walks in some measure by digging little round holes. It appears, by the dung that they drop upon the turf, that beetles are no inconsiderable part of their food. In June last I procured a litter of four or five young hedge-hogs, which appeared to be about five or six days old: they, I find, like puppies, are born blind, and could not see when they came to my hands. No doubt their spines are soft and flexible at the time of their birth, or else the poor dam would have but a bad time of it in the critical moment of parturition: but it is plain that they soon harden; for these little pigs had such stiff prickles on their backs and sides as would easily have fetched blood, had they not been handled with caution. Their spines are quite white at this age; and they have little hanging ears, which I do not remember to be discernible in the old ones. They can, in part, at this age draw their skin down over their faces; but are not able to contract themselves into a ball, as they do, for the sake of defence, when full grown. The reason, I suppose, is, because the curious muscle that enables the creature to roll itself up in a ball was not then arrived at its full tone and firmness. Hedge-hogs make a deep and warm *hybernaculum* with leaves and moss, in which they conceal themselves for the winter:* but I never could find that they stored in any winter provision, as some quadrupeds certainly do.

I have discovered an anecdote with respect to the fieldfare (*turdus pilaris*), which I think is particular enough: this bird, though it sits on trees in the day-time, and procures the greatest part of its food from white-thorn hedges; yea, moreover, builds on very high trees; as may be seen by the *fauna suecica*;* yet always appears with us to roost on the ground. They are seen to come in flocks just before it is dark, and to settle and nestle among the heath on our forest. And besides, the larkers, in dragging their nets by night, frequently catch them in the wheat-stubbles; while the bat-fowlers,* who take many red-wings in the hedges, never entangle any of this species. Why these birds, in the matter of roosting, should differ from all their congeners, and from themselves also with respect to their proceedings by day, is a fact for which I am by no means able to account.

I have somewhat* to inform you of concerning the moose-deer; but in general foreign animals fall seldom in my way: my little intelligence is confined to the narrow sphere of my own observations at home.

LETTER 28

Selborne, March 1770

On Michaelmas-day 1768 I managed to get a sight of the female moose belonging to the duke of Richmond,* at Goodwood; but was greatly disappointed, when I arrived at the spot, to find that it died, after having appeared in a languishing way for some time, on the morning before. However, understanding that it was not stripped, I proceeded to examine this rare quadruped: I found it in an old green-house, slung under the belly and chin by ropes, and in a standing posture; but, though it had been dead for so short a time, it was in so putrid a state that the stench was hardly supportable. The grand distinction between this deer, and any other species that I have ever met with, consisted in the strange length of its legs; on which it was tilted up much in the manner of the birds of the grallae order. I measured it, as they do an horse, and found that, from the ground to the wither,* it was just five feet four inches; which height answers exactly to sixteen hands, a growth that few horses arrive at: but then, with this length of legs, its neck was remarkably short, no more than twelve inches; so that, by straddling with one foot forward and the other backward, it grazed on the plain ground, with the greatest difficulty, between its legs: the ears were vast and lopping, and as long as the neck; the head was about twenty inches long, and ass-like; and had such a redundancy of upper lip as I never saw before, with huge nostrils. This lip, travellers say, is esteemed a dainty dish in North America. It is very reasonable to suppose that this creature supports itself chiefly by browsing of trees, and by wading after water plants; towards which way of livelihood the length of legs and great lip must contribute much. I have read somewhere that it delights in eating the *nymphaea*, or water-lily. From the fore-feet to the belly behind the shoulder it measured three feet and eight inches: the length of the legs before and behind consisted a great deal in the tibia, which was strangely long; but, in my haste to get out of the stench, I forgot to measure that joint exactly. Its scut seemed to be about an inch long; the colour was a grizzly black; the mane about four inches long; the fore-hoofs were upright and shapely, the hind flat and splayed. The spring before it was only two years old, so that most probably it was not then come to its growth. What a vast tall beast must a full grown

MOOSE DEER.

G. Stubbs pinx.   Mazell sculp.

stag be! I have been told some arrive at ten feet and an half! This poor creature had at first a female companion of the same species, which died the spring before. In the same garden was a young stag, or red deer, between whom and this moose it was hoped that there might have been a breed; but their inequality of height must have always been a bar to any commerce of the amorous kind. I should have been glad to have examined the teeth, tongue, lips, hoofs, &c. minutely; but the putrefaction precluded all farther curiosity. This animal, the keeper told me, seemed to enjoy itself best in the extreme frost of the former winter. In the house they shewed me the horn of a male moose, which had no front-antlers, but only a broad palm with some snags on the edge. The noble owner of the dead moose proposed to make a skeleton of her bones.

Please to let me hear if my female moose corresponds with that you saw; and whether you think still that the American moose and European elk are the same creature.*

I am,
With the greatest esteem, &c.

LETTER 29

Selborne, May 12, 1770

Dear Sir,

Last month we had such a series of cold turbulent weather, such a constant succession of frost, and snow, and hail, and tempest, that the regular migration or appearance of the summer birds was much interrupted. Some did not shew themselves (at least were not heard) till weeks after their usual time; as the black-cap and white-throat; and some have not been heard yet, as the grasshopper-lark and largest willow-wren. As to the fly-catcher, I have not seen it; it is indeed one of the latest, but should appear about this time: and yet, amidst all this meteorous strife and war of the elements, two swallows discovered themselves as long ago as the eleventh of April, in frost and snow; but they withdrew quickly, and were not visible again for many days. Housemartins, which are always more backward than swallows, were not observed till May came in.

OTTER.

Among the monogamous birds several are to be found, after pairing-time, single, and of each sex: but whether this state of celibacy is matter of choice or necessity, is not so easily discoverable. When the house-sparrows deprive my martins of their nests, as soon as I cause one to be shot, the other, be it cock or hen, presently procures a mate, and so for several times following.

I have known a dove-house infested by a pair of white owls, which made great havock among the young pigeons: one of the owls was shot as soon as possible; but the survivor readily found a mate, and the mischief went on. After some time the new pair were both destroyed, and the annoyance ceased.

Another instance I remember of a sportsman, whose zeal for the increase of his game being greater than his humanity, after pairing-time he always shot the cock-bird of every couple of partridges upon his grounds; supposing that the rivalry of many males interrupted the breed: he used to say, that, though he had widowed the same hen several times, yet he found she was still provided with a fresh paramour, that did not take her away from her usual haunt.

Again; I knew a lover of setting,* an old sportsman, who has often told me that soon after harvest he has frequently taken small coveys of partridges, consisting of cock-birds alone; these he pleasantly used to call old bachelors.

There is a propensity belonging to common house-cats that is very remarkable; I mean their violent fondness for fish, which appears to be their most favourite food: and yet nature in this instance seems to have planted in them an appetite that, unassisted, they know not how to gratify: for of all quadrupeds cats are the least disposed towards water; and will not, when they can avoid it, deign to wet a foot, much less to plunge into that element.

Quadrupeds that prey on fish are amphibious: such is the otter, which by nature is so well formed for diving, that it makes great havock among the inhabitants of the waters. Not supposing that we had any of those beasts in our shallow brooks, I was much pleased to see a male otter brought to me,* weighing twenty-one pounds, that had been shot on the bank of our stream below the Priory, where the rivulet divides the parish of Selborne from Harteley-wood.

LETTER 30

Selborne, Aug. 1, 1770

Dear Sir,

The French, I think, in general are strangely prolix in their natural history. What Linnaeus says with respect to insects holds good in every other branch: 'Verbositas praesentis saeculi, calamitas artis [The prolixity of the present age is a disaster for art].'

Pray how do you approve of Scopoli's new work? as I admire his *Entomologia*, I long to see it.*

I forgot to mention in my last letter (and had not room to insert in the former) that the male moose, in rutting time, swims from island to island, in the lakes and rivers of North-America, in pursuit of the females. My friend, the chaplain,* saw one killed in the water as it was on that errand in the river St Lawrence: it was a monstrous beast, he told me; but he did not take the dimensions.

When I was last in town our friend Mr Barrington most obligingly carried me to see many curious sights. As you were then writing to him about horns, he carried me to see many strange and wonderful specimens. There is, I remember, at Lord Pembroke's, at Wilton, an horn room furnished with more than thirty different pairs; but I have not seen that house lately.

Mr Barrington shewed me many astonishing collections of stuffed and living birds from all quarters of the world. After I had studied over the latter for a time, I remarked that every species almost that came from distant regions, such as South America, the coast of Guinea, &c. were thick-billed birds of the *loxia* and *fringilla* genera; and no *motacillae*, or *muscicapae*, were to be met with. When I came to consider, the reason was obvious enough; for the hard-billed birds subsist on seeds which are easily carried on board; while the soft-billed birds, which are supported by worms and insects, or, what is a *succedaneum* [substitute] for them, fresh raw meat, can meet with neither in long and tedious voyages. It is from this defect of food that our collections (curious as they are) are defective, and we are deprived of some of the most delicate and lively genera.

I am, &c.

LETTER 31

Selborne, Sept. 14, 1770

Dear Sir,

You saw, I find, the ring-ousels again among their native crags; and are farther assured that they continue resident in those cold regions the whole year. From whence then do our ring-ousels migrate so regularly every September, and make their appearance again, as if in their return, every April? They are more early this year than common, for some were seen at the usual hill on the fourth of this month.

An observing Devonshire gentleman* tells me that they frequent some parts of Dartmoor, and breed there; but leave those haunts about the end of September or beginning of October, and return again about the end of March.

Another intelligent person assures me that they breed in great abundance all over the Peak of Derby, and are called there Tor-ousels; withdraw in October and November, and return in spring. This information seems to throw some light on my new migration.

Scopoli's[1] new work (which I have just procured) has its merit in ascertaining many of the birds of the Tirol and Carniola. Monographers, come from whence they may, have, I think, fair pretence to challenge some regard and approbation from the lovers of natural history; for, as no man can alone investigate all the works of nature, these partial writers may, each in their department, be more accurate in their discoveries, and freer from errors, than more general writers; and so by degrees may pave the way to an universal correct natural history. Not that Scopoli is so circumstantial and attentive to the life and conversation* of his birds as I could wish: he advances some false facts; as when he says of the *hirundo urbica* that *'pullos extra nidum non nutrit* [it does not feed its young outside the nest]'. This assertion I know to be wrong from repeated observation this summer; for house-martins do feed their young flying, though it must be acknowledged not so commonly as the house-swallow; and the feat is done in so quick a manner as not to be perceptible to indifferent observers. He also advances some (I was going to say) improbable facts; as when he says of the woodcock that 'pullos rostro portat fugiens ab hoste [fleeing from an enemy, it carries its young in its beak]'. But candour forbids me to say absolutely that any fact is false, because I have never been witness to such a fact.* I have only to remark that the long unweildy bill of the woodcock is perhaps the worst adapted of any among the winged creation for such a feat of natural affection.

I am, &c.

LETTER 32

Selborne, October 19, 1770

Dear Sir,

After an ineffectual search in Linnaeus, Brisson, &c. I begin to suspect that I discern my brother's *hirundo hyberna* in Scopoli's new discovered *hirundo rupestris*, p. 167. His description of 'Supra murina, subtus albida; rectrices maculâ ovali albâ in latere interno; pedes nudi, nigri; rostrum nigrum; remiges obscuriores quam plumae dorsales; rectrices remigibus concolores; caudâ emarginatâ, nec forcipatâ

---

[1] *Annus Primus Historico-Naturalis.*

[mouse-coloured above, whitish below; tail-feathers with an oval white spot on the inner edge; feet black, with no feathers; bill black; flight feathers darker than the dorsal, with the tail feathers of the same tint; tail notched, not forked];' agrees very well with the bird in question: but when he comes to advance that it is 'statura *hirundinis urbicae* [the size of a house-martin]', and that 'definitio *hirundinis ripariae* Linnaei huic quoque convenit [the Linnaean definition of a sand-martin applies to this also]', he in some measure invalidates all he has said; at least he shews at once that he compares them to these species merely from memory: for I have compared the birds themselves, and find they differ widely in every circumstance of shape, size, and colour. However, as you will have a specimen, I shall be glad to hear what your judgment is in the matter.

Whether my brother is forestalled in his non-descript* or not, he will have the credit of first discovering that they spend their winters under the warm and sheltery shores of Gibraltar and Barbary.*

Scopoli's characters* of his ordines and genera are clean, just, and expressive, and much in the spirit of Linnaeus. These few remarks are the result of my first perusal of Scopoli's *Annus Primus*.

The bane of our science is the comparing one animal to the other by memory: for want of caution in this particular Scopoli falls into errors: he is not so full with regard to the manners* of his indigenous birds as might be wished, as you justly observe: his Latin is easy, elegant, and expressive, and very superior to Kramer's.[1]

I am pleased to see that my description of the moose corresponds so well with yours.

I am, &c.

LETTER 33

Selborne, Nov. 26, 1770

Dear Sir,

I was much pleased to see, among the collection of birds from Gibraltar, some of those short-winged English summer-birds of passage, concerning whose departure we have made so much inquiry.

[1] See his *Elenchus vegetabilium et animalium per Austriam inferiorem, &c.*

Now if these birds are found in Andalusia to migrate to and from Barbary, it may easily be supposed that those that come to us may migrate back to the continent, and spend their winters in some of the warmer parts of Europe. This is certain, that many soft-billed birds that come to Gibraltar appear there only in spring and autumn, seeming to advance in pairs towards the northward, for the sake of breeding during the summer months; and retiring in parties and broods towards the south at the decline of the year: so that the rock of Gibraltar is the great rendezvous, and place of observation, from whence they take their departure each way towards Europe or Africa. It is therefore no mean discovery, I think, to find that our small short-winged summer birds of passage are to be seen spring and autumn on the very skirts of Europe; it is a presumptive proof of their emigrations.

Scopoli seems to me to have found the *hirundo melba*, the great Gibraltar swift, in Tirol, without knowing it. For what is his *hirundo alpina* but the afore-mentioned bird in other words? Says he 'Omnia prioris' (meaning the swift); 'sed pectus album; paulo major priore [Everything similar to the former but the breast is white and the body a little larger].' I do not suppose this to be a new species. It is true also of the *melba*, that 'nidificat in excelsis Alpium rupibus [it nests on lofty peaks in the Alps]'. Vid. *Annum Primum.*

My Sussex friend,* a man of observation and good sense, but no naturalist, to whom I applied on account of the stone-curlew, *oedicnemus*, sends me the following account: 'In looking over my Naturalist's Journal for the month of April, I find the stone-curlews are first mentioned on the seventeenth and eighteenth, which date seems to me rather late. They live with us all the spring and summer, and at the beginning of autumn prepare to take leave by getting together in flocks. They seem to me a bird of passage that may travel into some dry hilly country south of us, probably Spain, because of the abundance of sheep-walks in that country; for they spend their summers with us in such districts. This conjecture I hazard, as I have never met with any one that has seen them in England in the winter. I believe they are not fond of going near the water, but feed on earthworms, that are common on sheep-walks and downs. They breed on fallows and lay-fields abounding with grey mossy flints, which much resemble their young in colour; among which they skulk and conceal themselves. They make no nest, but lay their eggs on the bare ground, producing in common but two at a time. There is reason to think their young run soon after they are hatched; and that the old ones do not feed them, but only lead them about at the time of feeding, which, for the most part, is in the night.' Thus far my friend.

In the manners of this bird you see there is something very analogous to the bustard, whom it also somewhat resembles in aspect and make, and in the structure of its feet.

For a long time I have desired my relation* to look out for these birds in Andalusia; and now he writes me word that, for the first time, he saw one dead in the market on the third of September.

When the *oedicnemus* flies it stretches out its legs straight behind, like an heron.

I am &c.

LETTER 34

Selborne, March 30, 1771

Dear Sir,

There is an insect with us, especially on chalky districts, which is very troublesome and teasing all the latter end of the summer, getting into people's skins, especially those of women and children, and raising tumours which itch intolerably. This animal (which we call an harvest bug) is very minute, scarce discernible to the naked eye; of a bright scarlet colour, and of the genus of *Acarus*. They are to be met with in gardens on kidneybeans, or any legumens; but prevail only in the hot months of summer. Warreners,* as some have assured me, are much infested by them on chalky downs; where these insects swarm sometimes to so infinite a degree as to discolour their nets, and to give them a reddish cast, while the men are so bitten as to be thrown into fevers.

There is a small long shining fly in these parts very troublesome to the housewife, by getting into the chimnies, and laying its eggs in the bacon while it is drying: these eggs produce maggots called jumpers, which, harbouring in the gammons and best parts of the hogs, eat down to the bone, and make great waste. This fly I suspect to be a variety of the *musca putris* of Linnaeus: it is to be seen in the summer in farm-kitchens on the bacon-racks and about the mantle-pieces, and on the ceilings.

The insect that infests turnips and many crops in the garden (destroying often whole fields while in their seedling leaves) is an animal that wants to be better known. The country people here call it the turnip-fly and black-dolphin; but I know it to be one of the coleoptera;* the '*chrysomela oleracea, saltatoria, femoribus posticis crassissimis.*' In very hot summers they abound to an amazing degree, and, as you walk in a field or in a garden, make a pattering like rain, by jumping on the leaves of the turnips or cabbages.

There is an *Oestrus,** known in these parts to every ploughboy; which, because it is omitted by Linnaeus, is also passed over by late writers; and that is the *curvicauda* of old Moufet, mentioned by Derham in his *Physico-theology*, p. 250: an insect worthy of remark for depositing its eggs as it flies in so dextrous a manner on the single hairs of the legs and flanks of grass-horses. But then Derham is mistaken when he advances that this *Oestrus* is the parent of that wonderful star-tailed maggot which he mentions afterwards; for more modern entomologists have discovered that singular production to be derived from the egg of the *musca chamaeleon*: see Geoffroy, t. 17, f. 4.*

A full history of noxious insects hurtful in the field, garden, and house, suggesting all the known and likely means of destroying them, would be allowed by the public to be a most useful and important work. What knowledge there is of this sort lies scattered, and wants to be collected; great improvements would soon follow of course. A knowledge of the properties, oeconomy, propagation, and in short of the life and conversation of these animals, is a necessary step to lead us to some method of preventing their depredations.

As far as I am a judge, nothing would recommend entomology more than some neat plates that should well express the generic distinctions of insects according to Linnaeus; for I am well assured that many people would study insects, could they set out with a more adequate notion of those distinctions than can be conveyed at first by words alone.

LETTER 35

Selborne, 1771

Dear Sir,

Happening to make a visit to my neighbour's peacocks, I could not help observing that the trains of those magnificent birds appear by no means to be their tails; those long feathers growing not from their uropygium,* but all up their backs. A range of short brown stiff feathers, about six inches long, fixed in the uropygium, is the real tail, and serves as the fulcrum to prop the train, which is long and top-heavy, when set on end. When the train is up, nothing appears of the bird before but its head and neck; but this would not be the case were those long feathers fixed only in the rump, as may be seen by the turkey-cock when in a strutting attitude. By a strong muscular vibration these birds can make the shafts of their long feathers clatter like the swords of a sword-dancer; they then trample very quick with their feet, and run backwards towards the females.

I should tell you that I have got an uncommon *calculus aegogropila,** taken out of the stomach of a fat ox; it is perfectly round, and about the size of a large Seville orange; such are, I think, usually flat.

LETTER 36

Sept. 1771

Dear Sir,

The summer through I have seen but two of that large species of bat which I call *vespertilio altivolans*, from its manner of feeding high in the air: I procured one of them, and found it to be a male; and made no doubt, as they accompanied together, that the other was a female: but, happening in an evening or two to procure the other likewise, I was somewhat disappointed, when it appeared to be also of the same sex. This circumstance, and the great scarcity of this sort, at least in these parts, occasions some suspicions in my mind whether it is really a species, or whether it may not be the male part of the more known

GREAT BAT.

species, one of which may supply many females; as is known to be the case in sheep, and some other quadrupeds. But this doubt can only be cleared by a farther examination, and some attention to the sex, of more specimens: all that I know at present is, that my two were amply furnished with the parts of generation much resembling those of a boar.

In the extent of their wings they measured fourteen inches and an half; and four inches and an half from the nose to the tip of the tail: their heads were large, their nostrils bilobated, their shoulders broad and muscular; and their whole bodies fleshy and plump. Nothing could be more sleek and soft than their fur, which was of a bright chesnut colour; their maws were full of food, but so macerated that the quality could not be distinguished; their livers, kidnies, and hearts, were large, and their bowels covered with fat. They weighed each, when entire, full one ounce and one drachm.* Within the ear there was somewhat of a peculiar structure that I did not understand perfectly; but refer it to the observation of the curious anatomist. These creatures sent forth a very rancid and offensive smell.*

LETTER 37

Selborne, 1771

Dear Sir,

On the twelfth of July I had a fair opportunity of contemplating the motions of the *caprimulgus*, or fern-owl,* as it was playing round a large oak that swarmed with *scarabaei solstitiales*, or fern-chafers.* The powers of its wing were wonderful, exceeding, if possible, the various evolutions* and quick turns of the swallow genus. But the circumstance that pleased me most was, that I saw it distinctly, more than once, put out its short leg while on the wing, and, by a bend of the head, deliver somewhat into its mouth. If it takes any part of its prey with its foot, as I have now the greatest reason to suppose it does these chafers, I no longer wonder at the use of its middle toe, which is curiously furnished with a serrated claw.

Swallows and martins, the bulk of them I mean, have forsaken us sooner this year than usual; for, on September the twenty-second, they rendezvoused in a neighbour's walnut-tree, where it seemed probable they had taken up their lodging for the night. At the dawn of the day, which was foggy, they arose all together in infinite numbers, occasioning such a rushing from the strokes of their wings against the hazy air, as might be heard to a considerable distance: since that no flock has appeared, only a few stragglers.

Some swifts staid late, till the twenty-second of August—a rare instance! for they usually withdraw within the first week.[1]

On September the twenty-fourth three or four ring-ousels appeared in my fields for the first time this season: how punctual are these visitors in their autumnal and spring migrations!

LETTER 38

Selborne, March 15, 1773

Dear Sir,

By my journal for last autumn it appears that the house-martins bred very late, and staid very late in these parts; for, on the first of

---

[1] See letter 53, to Mr Barrington.

October, I saw young martins in their nest nearly fledged; and again, on the twenty-first of October, we had at the next house a nest full of young martins just ready to fly; and the old ones were hawking for insects with great alertness. The next morning the brood forsook their nest, and were flying round the village. From this day I never saw one of the swallow kind till November the third; when twenty, or perhaps thirty, house-martins were playing all day long by the side of the hanging wood, and over my fields. Did these small weak birds, some of which were nestlings twelve days ago, shift their quarters at this late season of the year to the other side of the northern tropic? Or rather, is it not more probable that the next church, ruin, chalk-cliff, steep covert, or perhaps sandbank, lake or pool (as a more northern naturalist would say), may become their *hybernaculum*,* and afford them a ready and obvious retreat?

We now begin to expect our vernal migration of ring-ousels every week. Persons worthy of credit assure me that ring-ousels were seen at Christmas 1770 in the forest of Bere, on the southern verge of this county. Hence we may conclude that their migrations are only internal, and not extended to the continent southward, if they do at first come at all from the northern parts of this island only, and not from the north of Europe. Come from whence they will, it is plain, from the fearless disregard that they shew for men or guns, that they have been little accustomed to places of much resort. Navigators mention that in the Isle of Ascension, and other such desolate districts, birds are so little acquainted with the human form that they settle on men's shoulders; and have no more dread of a sailor than they would have of a goat that was grazing. A young man at Lewes, in Sussex, assured me that about seven years ago ring-ousels abounded so about that town in the autumn that he killed sixteen himself in one afternoon: he added further, that some had appeared since in every autumn; but he could not find that any had been observed before the season in which he shot so many. I myself have found these birds in little parties in the autumn cantoned* all along the Sussex downs, wherever there were shrubs and bushes, from Chichester to Lewes; particularly in the autumn of 1770.

I am, &c.

LETTER 39

Selborne, Nov. 9, 1773

Dear Sir,

As you desire me to send you such observations as may occur, I take the liberty of making the following remarks, that you may, according as you think me right or wrong, admit or reject what I here advance, in your intended new edition of the *British Zoology*.*

The osprey* was shot about a year ago at Frinsham-pond, a great lake, at about six miles from hence, while it was sitting on the handle of a plough and devouring a fish: it used to precipitate itself into the water, and so take its prey by surprise.

A great ash-coloured butcher-bird was shot last winter in Tisted-park, and a red backed butcher-bird at Selborne: they are *rarae aves* [rare birds] in this county.

Crows go in pairs the whole year round.

Cornish choughs abound, and breed on Beachy-head and on all the cliffs of the Sussex coast.

The common wild-pigeon, or stock-dove, is a bird of passage in the south of England, seldom appearing till towards the end of November; is usually the latest winter-bird of passage. Before our beechen woods were so much destroyed we had myriads of them, reaching in strings for a mile together as they went out in a morning to feed. They leave us early in spring; where do they breed?

The people of Hampshire and Sussex call the missel-bird the storm-cock, because it sings early in the spring in blowing showery weather; its song often commences with the year: with us it builds much in orchards.

A gentleman assures me he has taken the nests of ring-ousels on Dartmoor: they build in banks on the sides of streams.

Titlarks not only sing sweetly as they sit on trees, but also as they play and toy about on the wing; and particularly while they are descending, and sometimes as they stand on the ground.

Adanson's[1] testimony seems to me to be a very poor evidence that European swallows migrate during our winter to Senegal: he does not talk at all like an ornithologist; and probably saw only the swallows of

[1] Pennant's *British Zoology*, vol. 2, p. 242.

that country, which I know build within Governor O'Hara's hall against the roof. Had he known European swallows, would he not have mentioned the species?*

The house-swallow washes by dropping into the water as it flies: this species appears commonly about a week before the house-martin and about ten or twelve days before the swift.

In 1772 there were young house-martins in their nest till October the twenty-third.

The swift appears about ten or twelve days later than the house swallow: *viz.* about the twenty-fourth or twenty-sixth of April.

Whin-chats and stone-chatters stay with us the whole year.

Some wheat-ears continue with us the winter through.

Wagtails, all sorts, remain with us all the winter.

Bulfinches, when fed on hempseed, often become wholly black.

We have vast flocks of female chaffinches all the winter, with hardly any males among them.

When you say that in breeding-time the cock-snipes make a bleating noise, and I a drumming (perhaps I should have rather said an humming), I suspect we mean the same thing. However, while they are playing about on the wing they certainly make a loud piping with their mouths: but whether that bleating or humming is ventriloquous,* or proceeds from the motion of their wings, I cannot say; but this I know, that when this noise happens the bird is always descending, and his wings are violently agitated.

Soon after the lapwings have done breeding they congregate, and, leaving the moors and marshes, betake themselves to downs and sheep-walks.

Two years ago last spring the little auk was found alive and unhurt, but fluttering and unable to rise, in a lane a few miles from Alresford, where there is a great lake: it was kept awhile, but died.

I saw young teals taken alive in the ponds of Wolmer-forest in the beginning of July last, along with flappers, or young wild-ducks.

Speaking of the swift, that page says 'its drink the dew;'* whereas it should be 'it drinks on the wing;' for all the swallow kind sip their water as they sweep over the face of pools or rivers: like Virgil's bees, they drink flying; 'flumina summa libant [they sip from the surface of the stream. Virgil, *Georgics* 4]'. In this method of drinking perhaps this genus may be peculiar.

Of the sedge-bird be pleased to say it sings most part of the night;

its notes are hurrying, but not unpleasing, and imitative of several birds; as the sparrow, swallow, sky-lark. When it happens to be silent in the night, by throwing a stone or clod into the bushes where it sits you immediately set it a singing; or in other words, though it slumbers sometimes, yet as soon as it is awakened it reassumes its song.

LETTER 40

Selborne, Sept. 2, 1774

Dear Sir,

Before your letter arrived, and of my own accord, I had been remarking and comparing the tails of the male and female swallow, and this ere any young broods appeared; so that there was no danger of confounding the dams with their pulli:* and besides, as they were then always in pairs, and busied in the employ of nidification,* there could be no room for mistaking the sexes, nor the individuals of different chimnies* the one for the other. From all my observations, it constantly appeared that each sex has the long feathers in its tail that give it that forked shape; with this difference, that they are longer in the tail of the male than in that of the female.

Nightingales, when their young first come abroad, and are helpless, make a plaintive and a jarring noise; and also a snapping or cracking, pursuing people along the hedges as they walk: these last sounds seem intended for menace and defiance.

The grasshopper-lark chirps all night in the height of summer.

Swans turn white the second year, and breed the third.

Weasels prey on moles, as appears by their being sometimes caught in mole-traps.

Sparrow-hawks sometimes breed in old crows' nests, and the kestril in churches and ruins.

There are supposed to be two sorts of eels in the island of Ely. The threads sometimes discovered in eels are perhaps their young: the generation of eels is very dark and mysterious.*

Hen-harriers breed on the ground, and seem never to settle on trees.

When redstarts shake their tails they move them horizontally, as dogs do when they fawn: the tail of a wagtail, when in motion, bobs up and down like that of a jaded* horse.

Hedge-sparrows have a remarkable flirt* with their wings in breeding-time; as soon as frosty mornings come they make a very piping plaintive noise.

Many birds which become silent about Midsummer reassume their notes again in September; as the thrush, blackbird, woodlark, willow-wren, &c.; hence August is by much the most mute month, the spring, summer, and autumn through. Are birds induced to sing again because the temperament of autumn resembles that of spring?

Linnaeus ranges plants geographically; palms inhabit the tropics, grasses the temperate zones, and mosses and lichens the polar circles; no doubt animals may be classed in the same manner with propriety.

House-sparrows build under eaves in the spring; as the weather becomes hotter they get out for coolness, and nest in plum-trees and apple-trees. These birds have been known sometimes to build in rooks' nests, and sometimes in the forks of boughs under rooks' nests.

As my neighbour was housing a rick he observed that his dogs devoured all the little red mice that they could catch, but rejected the common mice; and that his cats ate the common mice, refusing the red.

Red-breasts sing all through the spring, summer, and autumn. The reason that they are called autumn songsters is, because in the two first seasons their voices are drowned and lost in the general chorus; in the latter their song becomes distinguishable. Many songsters of the autumn seem to be the young cock red-breasts of that year: notwithstanding the prejudices in their favour, they do much mischief in gardens to the summer-fruits.[1]

The titmouse, which early in February begins to make two quaint notes, like the whetting of a saw, is the marsh titmouse: the great titmouse sings with three cheerful joyous notes, and begins about the same time.

Wrens sing all the winter through, frost excepted.

House-martins came remarkably late this year both in Hampshire

---

[1] They eat also the berries of the ivy, the honey-suckle, and the *euonymus europaeus*, or spindle-tree.

and Devonshire: is this circumstance for or against either hiding or migration?

Most birds drink sipping at intervals; but pigeons take a long continued draught, like quadrupeds.

Notwithstanding what I have said in a former letter, no grey crows were ever known to breed on Dartmoor; it was my mistake.

The appearance and flying of the *scarabaeus solstitialis*, or fern-chafer, commence with the month of July, and cease about the end of it. These scarabs are the constant food of *caprimulgi*, or fern owls, through that period. They abound on the chalky downs and in some sandy districts, but not in the clays.

In the garden of the Black-bear inn in the town of Reading is a stream or canal running under the stables and out into the fields on the other side of the road: in this water are many carps, which lie rolling about in sight, being fed by travellers, who amuse themselves by tossing them bread: but as soon as the weather grows at all severe these fishes are no longer seen, because they retire under the stables, where they remain till the return of spring. Do they lie in a torpid state? if they do not, how are they supported?

The note of the white-throat, which is continually repeated, and often attended with odd gesticulations on the wing, is harsh and displeasing. These birds seem of a pugnacious disposition; for they sing with an erected crest and attitudes of rivalry and defiance; are shy and wild in breeding-time, avoiding neighbourhoods, and haunting lonely lanes and commons; nay even the very tops of the Sussex-downs, where there are bushes and covert; but in July and August they bring their broods into gardens and orchards, and make great havock among the summer-fruits.

The black-cap has in common a full, sweet, deep, loud, and wild pipe; yet that strain is of short continuance, and his motions are desultory; but when that bird sits calmly and engages in song in earnest, he pours forth very sweet, but inward melody, and expresses great variety of soft and gentle modulations, superior perhaps to those of any of our warblers, the nightingale excepted.

Black-caps mostly haunt orchards and gardens; while they warble their throats are wonderfully distended.

The song of the redstart is superior, though somewhat like that of the white-throat: some birds have a few more notes than others. Sitting very placidly on the top of a tall tree in a village, the cock sings

from morning to night: he affects neighbourhoods,* and avoids soli-
tude, and loves to build in orchards and about houses; with us he
perches on the vane of a tall maypole.

The fly-catcher is of all our summer birds the most mute and the
most familiar; it also appears the last of any. It builds in a vine, or a
sweetbriar,* against the wall of an house, or in the hole of a wall, or
on the end of a beam or plate, and often close to the post of a door
where people are going in and out all day long. This bird does not
make the least pretension to song, but uses a little inward wailing
note when it thinks its young in danger from cats or other annoyances: it breeds but once, and retires* early.

Selborne parish alone can and has exhibited at times more than
half the birds that are ever seen in all Sweden; the former has produced more than one hundred and twenty species, the latter only two
hundred and twenty-one. Let me add also that it has shewn near half
the species that were ever known in Great-Britain.[1]

On a retrospect, I observe that my long letter carries with it a
quaint and magisterial air, and is very sententious; but, when I recollect that you requested stricture and anecdote, I hope you will pardon
the didactic manner for the sake of the information it may happen
to contain.

LETTER 41*

It is matter of curious inquiry to trace out how those species of soft-
billed birds, that continue with us the winter through, subsist during
the dead months. The imbecility* of birds seems not to be the only
reason why they shun the rigour of our winters; for the robust wryneck (so much resembling the hardy race of woodpeckers) migrates,
while the feeble little golden-crowned wren, that shadow of a bird,
braves our severest frosts without availing himself of houses or villages, to which most of our winter-birds crowd in distressful seasons,
while this keeps aloof in fields and woods; but perhaps this may be the
reason why they may often perish, and why they are almost as rare as
any bird we know.

I have no reason to doubt but that the soft-billed birds, which

[1] Sweden 221, Great-Britain 252 species.

winter with us, subsist chiefly on insects in their aurelia* state. All the
species of wagtails in severe weather haunt shallow streams near their
spring-heads, where they never freeze; and, by wading, pick out the
aurelias of the genus of *Phryganeae** &c.[1]

Hedge-sparrows frequent sinks* and gutters in hard weather,
where they pick up crumbs and other sweepings: and in mild weather
they procure worms, which are stirring every month in the year, as
any one may see that will only be at the trouble of taking a candle to a
grass-plot on any mild winter's night. Redbreasts and wrens in the
winter haunt out-houses, stables, and barns, where they find spiders
and flies that have laid themselves up during the cold season. But the
grand support of the soft-billed birds in winter is that infinite profusion of aureliae of the lepidoptera *ordo*,* which is fastened to the twigs
of trees and their trunks; to the pales and walls of gardens and buildings; and is found in every cranny and cleft of rock or rubbish, and
even in the ground itself.

Every species of titmouse winters with us; they have what I call a
kind of intermediate bill between the hard and the soft, between the
Linnaean genera of *fringilla* and *motacilla*. One species alone spends
its whole time in the woods and fields, never retreating for succour in
the severest seasons to houses and neighbourhoods; and that is the
delicate long-tailed titmouse, which is almost as minute as the golden-
crowned wren: but the blue titmouse, or nun (*parus caeruleus*), the
cole-mouse (*parus ater*), the great black-headed titmouse (*fringillago*),
and the marsh titmouse (*parus palustris*), all resort, at times, to buildings; and in hard weather particularly. The great titmouse, driven by
stress of weather, much frequents houses; and, in deep snows, I have
seen this bird, while it hung with its back downwards (to my no small
delight and admiration), draw straws lengthwise from out the eaves
of thatched houses, in order to pull out the flies that were concealed
between them, and that in such numbers that they quite defaced the
thatch, and gave it a ragged appearance.

The blue titmouse, or nun, is a great frequenter of houses, and a
general devourer. Besides insects, it is very fond of flesh; for it frequently picks bones on dunghills: it is a vast admirer of suet, and
haunts butchers' shops. When a boy, I have known twenty in a morning caught with snap mouse-traps, baited with tallow or suet. It will

[1] See Derham's *Physico-theology*, p. 235.

also pick holes in apples left on the ground, and be well entertained with the seeds on the head of a sun-flower. The blue, marsh, and great titmice will, in very severe weather, carry away barley and oat straws from the sides of ricks.

How the wheat-ear and whin-chat support themselves in winter cannot be so easily ascertained, since they spend their time on wild heaths and warrens*; the former especially, where there are stone quarries: most probably it is that their maintenance arises from the aureliae of the lepidoptera *ordo*, which furnish them with a plentiful table in the wilderness.

I am, &c.

LETTER 42

Selborne, March 9, 1775

Dear Sir,

Some future faunist,* a man of fortune, will, I hope, extend his visits to the kingdom of Ireland; a new field, and a country little known to the naturalist. He will not, it is to be wished, undertake that tour unaccompanied by a botanist, because the mountains have scarcely been sufficiently examined; and the southerly counties of so mild an island may possibly afford some plants little to be expected within the British dominions. A person of a thinking turn of mind will draw many just remarks from the modern improvements of that country, both in arts and agriculture, where premiums obtained long before they were heard of with us.* The manners of the wild natives, their superstitions, their prejudices, their sordid way of life, will extort from him many useful reflections. He should also take with him an able draughtsman; for he must by no means pass over the noble castles and seats, the extensive and picturesque lakes and waterfalls, and the lofty stupendous mountains, so little known, and so engaging to the imagination when described and exhibited in a lively manner: such a work would be well received.

As I have seen no modern map of Scotland, I cannot pretend to say how accurate or particular any such may be; but this I know, that the best old maps of that kingdom are very defective.

The great obvious defect that I have remarked in all maps of

Scotland that have fallen in my way is, a want of a coloured line, or stroke, that shall exactly define the just limits of that district called The Highlands. Moreover, all the great avenues to that mountainous and romantic country want to be well distinguished. The military roads formed by general Wade are so great and Roman-like an undertaking that they well merit attention. My old map, Moll's Map, takes notice of Fort William; but could not mention the other forts that have been erected long since: therefore a good representation of the chain of forts should not be omitted.*

The celebrated zigzag up the Coryarich* must not be passed over. Moll takes notice of Hamilton and Drumlanrig, and such capital houses; but a new survey, no doubt, should represent every seat and castle remarkable for any great event, or celebrated for its paintings, &c. Lord Breadalbane's seat and beautiful policy* are too curious and extraordinary to be omitted.

The seat of the Earl of Eglintoun, near Glasgow, is worthy of notice. The pine-plantations of that nobleman are very grand and extensive indeed.

I am, &c.

LETTER 43

A pair of honey-buzzards, *buteo apivorus, sive vespivorus Raii*, built them a large shallow nest, composed of twigs and lined with dead beechen leaves, upon a tall slender beech near the middle of Selborne-hanger, in the summer of 1780. In the middle of the month of June a bold boy climbed this tree, though standing on so steep and dizzy a situation, and brought down an egg, the only one in the nest, which had been sat on for some time, and contained the embrio of a young bird. The egg was smaller, and not so round as those of the common buzzard; was dotted at each end with small red spots, and surrounded in the middle with a broad bloody zone.

The hen-bird was shot, and answered exactly to Mr Ray's description of that species; had a black cere,* short thick legs, and a long tail. When on the wing this species may be easily distinguished from the common buzzard by its hawk-like appearance, small head, wings not so blunt, and longer tail. This specimen contained in its craw some limbs of frogs and many grey snails without shells. The irides of the eyes of this bird were of a beautiful bright yellow colour.

About the tenth of July in the same summer a pair of sparrow-hawks bred in an old crow's nest on a low beech in the same hanger; and as their brood, which was numerous, began to grow up, became so daring and ravenous, that they were a terror to all the dames in the village that had chickens or ducklings under their care. A boy climbed the tree, and found the young so fledged that they all escaped from him; but discovered that a good house had been kept: the larder was well-stored with provisions; for he brought down a young blackbird, jay, and house-martin, all clean picked, and some half devoured. The old birds had been observed to make sad havock for some days among the new-flown swallows and martins, which, being but lately out of their nests, had not acquired those powers and command of wing that enable them, when more mature, to set such enemies at defiance.

LETTER 44

Selborne, Nov. 30, 1780

Dear Sir,

Every incident that occasions a renewal of our correspondence will ever be pleasing and agreeable to me.

As to the wild wood-pigeon, the *oenas*, or *vinago*, of Ray, I am much of your mind; and see no reason for making it the origin of the common house-dove: but suppose those that have advanced that opinion may have been misled by another appellation, often given to the *oenas*, which is that of stock-dove.

Unless the stock-dove in the winter varies greatly in manners from itself in summer, no species seems more unlikely to be domesticated, and to make an house-dove. We very rarely see the latter settle on trees at all, nor does it ever haunt the woods; but the former, as long as it stays with us, from November perhaps to February, lives the same wild life with the ring-dove, *palumbus torquatus*; frequents coppices and groves, supports itself chiefly by mast, and delights to roost in the tall-est beeches. Could it be known in what manner stock-doves build, the doubt would be settled with me at once, provided they construct their nests on trees, like the ring-dove, as I much suspect they do.

You received, you say, last spring a stock-dove from Sussex; and are informed that they sometimes breed in that country. But why did not your correspondent determine the place of its nidification, whether on rocks, cliffs, or trees? If he was not an adroit ornithologist I should doubt the fact, because people with us perpetually confound the stock-dove with the ring-dove.

For my own part, I readily concur with you in supposing that house-doves are derived from the small blue rock-pigeon, for many reasons. In the first place the wild stock-dove is manifestly larger than the common house-dove, against the usual rule of domestication, which generally enlarges the breed. Again, those two remarkable black spots on the remiges* of each wing of the stock-dove, which are so characteristic of the species, would not, one should think, be totally lost by its being reclaimed; but would often break out among its descendants. But what is worth an hundred arguments is, the instance you give in Sir Roger Mostyn's house-doves in Caernarvonshire;* which, though tempted by plenty of food and gentle treatment, can never be prevailed on to inhabit their cote for any time; but, as soon as they begin to breed, betake themselves to the fastnesses* of Ormshead, and deposit their young in safety amidst the inaccessible caverns, and precipices of that stupendous promontory.

> Naturam expellas furcâ . . . tamen usque recurret.
>
> [Nature can be driven out with a pitchfork . . . but
>   she will always return.]
>
> [Horace, *Epistles* I. 10]

I have consulted a sportsman, now in his seventy-eighth year, who tells me that fifty or sixty years back, when the beechen woods were much more extensive than at present, the number of wood-pigeons was astonishing; that he has often killed near twenty in a day; and that with a long wild-fowl piece he has shot seven or eight at a time on the wing as they came wheeling over his head: he moreover adds, which I was not aware of, that often there were among them little parties of small blue doves, which he calls rockiers. The food of these numberless emigrants was beech-mast and some acorns; and particularly barley, which they collected in the stubbles. But of late years, since the vast increase of turnips, that vegetable has furnished a great part of their support in hard weather; and the holes they pick in these roots greatly damage the crop. From this food their flesh has contracted a rancidness which occasions them to be rejected by nicer judges of eating, who thought them before a delicate dish. They were shot not only as they were feeding in the fields, and especially in snowy weather, but also at the close of the evening, by men who lay in ambush among the woods and groves to kill them as they came in to roost.[1] These are the principal circumstances relating to this wonderful internal migration, which with us takes place towards the end of November, and ceases early in the spring. Last winter we had in Selborne high wood about an hundred of these doves; but in former times the flocks were so vast, not only with us but all the district round, that on mornings and evenings they traversed the air, like rooks, in strings, reaching for a mile together. When they thus rendezvoused here by thousands, if they happened to be suddenly roused from their roost-trees on an evening,

> Their rising all at once was like the sound*
> Of thunder heard remote. . . .
>
> [Milton, *Paradise Lost*, II. 476–7]

It will by no means be foreign to the present purpose to add, that I had a relation in this neighbourhood who made it a practice, for a time, whenever he could procure the eggs of a ring-dove, to place them under a pair of doves that were sitting in his own pigeon-house; hoping thereby, if he could bring about a coalition, to enlarge his breed, and teach his own doves to beat out into the woods and to support themselves by mast:* the plan was plausible, but something always interrupted the success; for though the birds were usually hatched, and sometimes grew to half their size, yet none ever arrived at maturity. I myself have seen these foundlings in their nest displaying a strange ferocity of nature, so as scarcely to bear to be looked at, and snapping with their bills by way of menace. In short, they always died, perhaps for want of proper sustenance: but the owner thought that by their fierce and wild demeanour they frighted their foster-mothers, and so were starved.

Virgil, as a familiar occurrence, by way of simile, describes a dove haunting the cavern of a rock in such engaging numbers, that I cannot refrain from quoting the passage: and John Dryden has rendered it so

---

[1] Some old sportsmen say that the main part of these flocks used to withdraw as soon as the heavy Christmas frosts were over.

happily in our language, that without farther excuse I shall add his translation also.

> Qualis speluncâ subitò commota Columba,
> Cui domus, et dulces latebroso in pumice nidi,
> Fertur in arva volans, plausumque exterrita pennis
> Dat tecto ingentem—mox aere lapsa quieto,
> Radit iter liquidum, celeres neque commovet alas.

> As when a dove her rocky hold forsakes,
> Rous'd, in a fright her sounding wings she shakes;
> The cavern rings with clattering:—out she flies,
> And leaves her callow care, and cleaves the skies:
> At first she flutters:—but at length she springs
> To smoother flight, and shoots upon her wings.

[Virgil, *Aeneid* 5]

I am, &c.

LETTERS
TO THE
HONOURABLE DAINES BARRINGTON

LETTER 1

Selborne, June 30, 1769

Dear Sir,

When I was in town last month* I partly engaged that I would sometime do myself the honour to write to you on the subject of natural history: and I am the more ready to fulfil my promise, because I see you are a gentleman of great candour, and one that will make allowances; especially where the writer professes to be an out-door naturalist, one that takes his observations from the subject itself, and not from the writings of others.

The following is a LIST of the SUMMER BIRDS of PASSAGE which I have discovered in this neighbourhood, ranged somewhat in the order in which they appear:

		RAII NOMINA	USUALLY APPEARS ABOUT
1.	Wryneck	*Jynx, sive torquilla*	The middle of March: harsh note.
2.	Smallest willow-wren	*Regulus non cristatus*	March 23: chirps till September.
3.	Swallow	*Hirundo domestica*	April 13.
4.	Martin	*Hirundo rustica*	Ditto.
5.	Sand-martin	*Hirundo riparia*	Ditto.
6.	Black-cap	*Atricapilla*	Ditto: a sweet wild note.
7.	Nightingale	*Luscinia*	Beginning of April.
8.	Cuckoo	*Cuculus*	Middle of April.
9.	Middle willow-wren	*Regulus non cristatus*	Ditto: a sweet plaintive note.
10.	White-throat	*Ficedulae affinis*	Ditto: mean note; sings on till September.
11.	Red-start	*Ruticilla*	Ditto: more agreeable song.
12.	Stone-curlew	*Oedicnemus*	End of March: loud nocturnal whistle.
13.	Turtle-dove	*Turtur*	
14.	Grasshopper-lark	*Alauda minima locustae voce*	Middle April: a small sibilous note, till the end of July.
15.	Swift	*Hirundo apus*	About April 27.
16.	Less reed-sparrow	*Passer arundinaceus minor*	A sweet polyglot, but hurrying: it has the notes of many birds.
17.	Land-rail	*Ortyigometra*	A loud harsh note, crex, crex.

18. Largest willow-wren	*Regulus non cristatus*	Cantat voce stridulâ locustae [sings with the rasping sound of a grasshopper]; end of April, on the tops of high beeches.
19. Goatsucker, or fern-owl	*Caprimulgus*	Beginning of May; chatters by night with a singular noise.
20. Fly-catcher	*Stoparola*	May 12. A very mute bird: This is the latest summer bird of passage.

This assemblage of curious and amusing* birds belongs to ten several genera of the Linnaean system; and are all of the *ordo* of passeres save the *jynx* and *cuculus*, which are picae, and the *charadrius* (*oedicnemus*) and *rallus* (*ortygometra*) which are grallae.

These birds, as they stand numerically,* belong to the following Linnaean genera:

1:	*Jynx.*	13:	*Columba.*
2, 6, 7, 9, 10, 11, 16, 18:	*Motacilla.*	17:	*Rallus.*
3, 4, 5, 15:	*Hirundo.*	19:	*Caprimulgus.*
8:	*Cuculus.*	14:	*Alauda.*
12:	*Charadrius.*	20:	*Muscicapa.*

Most soft-billed birds live on insects, and not on grain and seeds; and therefore at the end of summer they retire: but the following soft-billed birds, though insect-eaters, stay with us the year round:

RAII NOMINA

Redbreast	*Rubecula*	These frequent houses; and haunt outbuildings in the winter: eat spiders.
Wren	*Passer troglodytes*	
Hedge-sparrow	*Curruca*	Haunt sinks for crumbs and other sweepings.
White-wagtail	*Motacilla alba*	These frequent shallow rivulets near the spring heads, where they never freeze: eat the aureliae of *Phryganea*. The smallest birds that walk.
Yellow wagtail	*Motacilla flava*	
Grey wagtail	*Motacilla cinerea*	

Wheat-ear	*Oenanthe*	{ Some of these are to be seen with us the winter through.
Whin-chat	*Oenanthe secunda.*	
Stone-chatter	*Oenanthe tertia.*	
Golden-crowned wren	*Regulus cristatus*	{ This is the smallest British bird: haunts the tops of tall trees; stays the winter through.

A LIST of the WINTER BIRDS of PASSAGE round this neighbourhood, ranged somewhat in the order in which they appear:

RAII NOMINA

1. Ring-ousel	*Merula torquata*	{ This is a new migration, which I have lately discovered about Michaelmas week, and again about the fourteenth of March.
2. Redwing	*Turdus iliacus*	About old Michaelmas.*
3. Fieldfare	*Turdus pilaris*	Though a percher by day, roosts on the ground.
4. Royston-crow	*Cornix cinerea*	Most frequent on downs.
5. Woodcock	*Scolopax*	Appears about old Michaelmas.
6. Snipe	*Gallinago minor*	Some snipes constantly breed with us.
7. Jack-snipe	*Gallinago minima.*	
8. Wood-pigeon	*Oenas*	Seldom appears till late: not in such plenty as formerly.
9. Wild-swan	*Cygnus ferus*	On some large waters.
10. Wild-goose	*Anser ferus.*	
11. Wild-duck	*Anas torquata minor*	
12. Pochard	*Anas fera fusca*	
13. Wigeon	*Penelope*	} On our lakes and streams.
14. Teal, breeds with us in Wolmer-forest	*Querquedula*	
15. Cross-beak	*Coccothraustes*	{ These are only wanderers that appear occasionally, and are not observant of any regular migration.
16. Gross-bill	*Loxia*	
17. Silk-tail	*Garrulus bohemicus*	

These birds, as they stand numerically, belong to the following Linnaean genera:

1, 2, 3:	*Turdus.*	9, 10, 11, 12, 13, 14:	*Anas.*
4:	*Corvus.*	15, 16:	*Loxia.*
5, 6, 7:	*Scolopax.*	17:	*Ampelis.*
8:	*Columba.*		

Birds that sing in the night are but few.

Nightingale	*Luscinia*	'In shadiest covert hid.' Milton [Paradise Lost, III. 39].
Woodlark	*Alauda arborea*	Suspended in mid air.
Less reed-sparrow	*Passer arundinaceus minor*	Among reeds and willows.

I should now proceed to such birds as continue to sing after Midsummer, but, as they are rather numerous, they would exceed the bounds of this paper: besides, as this is now the season for remarking on that subject, I am willing to repeat my observations on some birds concerning the continuation of whose song I seem at present to have some doubt.

I am, &c.

LETTER 2

Selborne, Nov. 2, 1769

Dear Sir,

When I did myself the honour to write to you about the end of last June on the subject of natural history, I sent you a list of the summer-birds of passage which I have observed in this neighbourhood; and also a list of the winter-birds of passage: I mentioned besides those soft-billed birds that stay with us the winter through in the south of England, and those that are remarkable for singing in the night.

According to my proposal, I shall now proceed to such birds (sing-ing birds strictly so called) as continue in full song till after Midsummer; and shall range them somewhat in the order in which they first begin to open as the spring advances.

RAII NOMINA

1. Wood-lark	*Alauda arborea*	In January, and continues to sing through all the summer and autumn.
2. Song-thrush	*Turdus simpliciter dictus*	In February and on to August, reassume their song in autumn.
3. Wren	*Passer troglodytes*	All the year, hard frost excepted.
4. Redbreast	*Rubecula*	Ditto.
5. Hedge-sparrow	*Curruca*	Early in February to July the 10th.
6. Yellowhammer	*Emberiza flava*	Early in February, and on through July to August the 21st.
7. Skylark	*Alauda vulgaris*	In February, and on to October.
8. Swallow	*Hirundo domestica*	From April to September.
9. Black-cap	*Atricapilla*	Beginning of April to July 13th.
10. Titlark	*Alauda pratorum*	From middle of April to July the 16th.
11. Blackbird	*Merula vulgaris*	Sometimes in February and March, and so on to July the twenty-third; reassumes in autumn.
12. White-throat	*Ficedulae affinis*	In April, and on to July 23.
13. Goldfinch	*Carduelis*	April, and through to September 16.
14. Greenfinch	*Chloris*	On to July and August 2.
15. Less reed-sparrow	*Passer arundinaceus minor*	May, on to beginning of July.
16. Common linnet	*Linaria vulgaris*	Breeds and whistles on till August; reassumes its note when they begin to congregate in October, and again early before the flocks separate.

Birds that cease to be in full song, and are usually silent at or before Midsummer:

17. Middle willow-wren	*Regulus non cristatus*	Middle of June: begins in April.
18. Redstart	*Ruticilla*	Ditto: begins in May.
19. Chaffinch	*Fringilla*	Beginning of June: sings first in February.
20. Nightingale	*Luscinia*	Middle of June: sings first in April.

Birds that sing for a short time, and very early in the spring:

21. Missel-bird	*Turdus viscivorus*	January the 2nd, 1770, in February. Is called in Hampshire and Sussex the storm-cock, because its song is supposed to forebode windy wet weather: is the largest singing bird we have.
22. Great titmouse, or ox eye	*Fringillago*	In February, March, April: reassumes for a short time in September.

Birds that have somewhat of a note or song, and yet are hardly to be called singing birds:

RAII NOMINA

23. Golden-crowned wren	*Regulus cristatus*	Its note as minute as its person; frequents the tops of high oaks and firs; the smallest British bird.
24. Marsh-titmouse	*Parus palustris*	Haunts great woods: two harsh sharp notes.
25. Small willow-wren	*Regulus non cristatus*	Sings in March and on to September.
26. Largest ditto	*Ditto*	Cantat voce stridulâ locustae [sings with the rasping sound of a grasshopper]: from end of April to August.
27. Grasshopper-lark	*Alauda minima voce locustae*	Chirps all night, from the middle of April to the end of July.
28. Martin	*Hirundo agrestis*	All the breeding time; from May to September.
29. Bullfinch	*Pyrrhula*	
30. Bunting	*Emberiza alba*	From the end of January to July.

All singing birds, and those that have any pretensions to song, not only in Britain, but perhaps the world through, come under the Linnaean *ordo* of passeres.

The above-mentioned birds, as they stand numerically, belong to the following Linnaean genera.

1, 7, 10, 27:	*Alauda.*	8, 28:	*Hirundo.*
2, 11, 21:	*Turdus.*	13, 16, 19:	*Fringilla.*
3, 4, 5, 9, 12, 15, 17, 18, 20, 23, 25, 26:	*Motacilla.*	22, 24:	*Parus.*
6, 30:	*Emberiza.*	14, 29:	*Loxia.*

Birds that sing as they fly are but few.

	RAII NOMINA	
Skylark	*Alauda vulgaris*	Rising, suspended, and falling.
Titlark	*Alauda pratorum*	In its descent; also sitting on trees, and walking on the ground.
Woodlark	*Alauda arborea*	Suspended; in hot summer nights all night long.
Blackbird	*Merula*	Sometimes from bush to bush.
White-throat	*Ficedulae affinis*	Uses when singing on the wing odd jerks and gesticulations.
Swallow	*Hirundo domestica*	In soft sunny weather.
Wren	*Passer troglodytes*	Sometimes from bush to bush.

Birds that breed most early in these parts:

Raven	*Corvus*	Hatches in February and March.
Song-thrush	*Turdus*	In March.
Blackbird	*Merula*	In March.
Rook	*Cornix frugilega*	Builds the beginning of March.
Woodlark	*Alauda arborea*	Hatches in April.
Ring-dove	*Palumbus torquatus*	Lays the beginning of April.

All birds that continue in full song till after Midsummer appear to me to breed more than once.

Most kinds of birds seem to me to be wild and shy somewhat in proportion to their bulk; I mean in this island, where they are much pursued and annoyed: but in Ascension Island, and many other desolate places, mariners have found fowls so unacquainted with an human figure, that they would stand still to be taken; as is the case with boobies,* &c. As an example of what is advanced, I remark that the golden-crested wren (the smallest British bird) will stand unconcerned till you come within three or four yards of it, while the bustard (*otis*), the largest British land fowl, does not care to admit a person within so many furlongs.

I am, &c.

LETTER 3

Selborne, Jan. 15, 1770

Dear Sir,

It was no small matter of satisfaction to me to find that you were not displeased with my little *methodus* of birds. If there was any merit in the sketch, it must be owing to its punctuality. For many months I carried a list in my pocket of the birds that were to be remarked, and, as I rode or walked about my business, I noted each day the continuance or omission of each bird's song; so that I am as sure of the certainty of my facts as a man can be of any transaction whatsoever.

I shall now proceed to answer the several queries which you put in your two obliging letters, in the best manner that I am able. Perhaps Eastwick, and its environs, where you heard so very few birds, is not a woodland country, and therefore not stocked with such songsters. If you will cast your eye on my last letter, you will find that many species continued to warble after the beginning of July.

The titlark and yellowhammer breed late, the latter very late; and therefore it is no wonder that they protract their song: for I lay it down as a maxim in ornithology, that as long as there is any incuba-tion going on there is music. As to the redbreast and wren, it is well known to the most incurious observer that they whistle the year round, hard frost excepted; especially the latter.

It was not in my power to procure you a black-cap, or a less reed-sparrow, or sedge-bird, alive. As the first is undoubtedly, and the last, as far as I can yet see, a summer bird of passage, they would require more nice and curious management in a cage than I should be able to give them: they are both distinguished songsters. The note of the former has such a wild sweetness that it always brings to my mind those lines in a song in *As You Like It*.

> And tune his merry note
> Unto the wild bird's throat.

Shakespeare*

The latter has a surprising variety of notes resembling the song of several other birds; but then it has also an hurrying manner, not at all to its advantage: it is notwithstanding a delicate polyglot.*

SNIPE.

It is new to me that titlarks in cages sing in the night; perhaps only caged birds do so. I once knew a tame redbreast in a cage that always sang as long as candles were in the room; but in their wild state no one supposes they sing in the night.

I should be almost ready to doubt the fact, that there are to be seen much fewer birds in July than in any former month, notwithstanding so many young are hatched daily. Sure I am that it is far otherwise with respect to the swallow tribe, which increases prodigiously as the summer advances: and I saw, at the time mentioned, many hundreds of young wagtails on the banks of the Cherwell,* which almost covered the meadows. If the matter appears as you say in the other species, may it not be owing to the dams being engaged in incubation, while the young are concealed by the leaves?

Many times have I had the curiosity to open the stomachs of woodcocks and snipes; but nothing ever occurred that helped to explain to

me what their subsistence might be: all that I could ever find was a
soft mucus, among which lay many pellucid small gravels.

I am, &c.

LETTER 4

Selborne, Feb. 19, 1770

Dear Sir,

Your observation that 'the cuckoo does not deposit its egg indiscriminately in the nest of the first bird that comes in its way, but
probably looks out a nurse in some degree congenerous, with whom
to intrust its young,' is perfectly new to me; and struck me so forcibly, that I naturally fell into a train of thought that led me to consider whether the fact was so, and what reason there was for it. When
I came to recollect and inquire, I could not find that any cuckoo had
ever been seen in these parts, except in the nest of the wagtail, the
hedge-sparrow, the titlark, the white-throat, and the redbreast, all
soft-billed insectivorous birds. The excellent Mr Willughby mentions the nest of the *palumbus* (ring-dove), and of the *fringilla* (chaffinch), birds that subsist on acorns and grains, and such hard food:
but then he does not mention them as of his own knowledge; but says
afterwards that he saw himself a wagtail feeding a cuckoo. It appears
hardly possible that a soft-billed bird should subsist on the same food
with the hard-billed: for the former have thin membranaceous stomachs suited to their soft food; while the latter, the granivorous tribe,
have strong muscular gizzards, which, like mills, grind, by the help of
small gravels and pebbles, what is swallowed. This proceeding of the
cuckoo, of dropping its eggs as it were by chance, is such a monstrous
outrage on maternal affection, one of the first great dictates of nature;
and such a violence on instinct; that, had it only been related of a bird
in the Brasils, or Peru, it would never have merited our belief. But
yet, should it farther appear that this simple bird, when divested of
that natural στοργή [affection] that seems to raise the kind in general
above themselves, and inspire them with extraordinary degrees of
cunning and address, may be still endued with a more enlarged faculty of discerning what species are suitable and congenerous nursing-
mothers for its disregarded eggs and young, and may deposit them

only under *their* care, this would be adding wonder to wonder, and instancing, in a fresh manner, that the methods of Providence are not subjected to any mode or rule, but astonish us in new lights, and in various and changeable appearances.

What was said by a very ancient and sublime writer concerning the defect of natural affection in the ostrich,* may be well applied to the bird we are talking of:

She is hardened against her young ones, as though they were not hers:

Because God hath deprived her of wisdom, neither hath he imparted to her understanding.[1]

Query. Does each female cuckoo lay but one egg in a season, or does she drop several in different nests according as opportunity offers?

I am, &c.

LETTER 5

Selborne, April 12, 1770

Dear Sir,

I heard many birds of several species sing last year after Midsummer; enough to prove that the summer solstice is not the period that puts a stop to the music of the woods. The yellowhammer no doubt persists with more steadiness than any other; but the woodlark, the wren, the redbreast, the swallow, the white-throat, the goldfinch, the common linnet, are all undoubted instances of the truth of what I advanced.

If this severe season does not interrupt the regularity of the summer migrations, the blackcap will be here in two or three days. I wish it was in my power to procure you one of those songsters; but I am no birdcatcher; and so little used to birds in a cage, that I fear if I had one it would soon die for want of skill in feeding.

Was your reed-sparrow, which you kept in a cage, the thick-billed reed-sparrow of the *Zoology*, p. 320; or was it the less reed-sparrow of Ray, the sedge-bird of Mr Pennant's last publication, p. 16?*

As to the matter of long-billed birds growing fatter in moderate frosts, I have no doubt within myself what should be the reason.

[1] Job 39: 16, 17.

The thriving at those times appears to me to arise altogether from the gentle check which the cold throws upon insensible perspiration. The case is just the same with blackbirds, &c.; and farmers and warreners observe, the first, that their hogs fat more kindly at such times, and the latter that their rabbits are never in such good case as in a gentle frost. But when frosts are severe, and of long continuance, the case is soon altered; for then a want of food soon overbalances the repletion occasioned by a checked perspiration. I have observed, moreover, that some human constitutions are more inclined to plumpness in winter than in summer.*

When birds come to suffer by severe frost, I find that the first that fail and die are the redwing-fieldfares, and then the song-thrushes.

You wonder, with good reason, that the hedge-sparrows, &c. can be induced at all to sit on the egg of the cuckoo without being scandalized at the vast disproportioned size of the supposititious* egg; but the brute creation, I suppose, have very little idea of size, colour, or number. For the common hen, I know, when the fury of incubation is on her, will sit on a single shapeless stone instead of a nest full of eggs that have been withdrawn: and, moreover, a hen-turkey, in the same circumstances, would sit on in the empty nest till she perished with hunger.

I think the matter might easily be determined whether a cuckoo lays one or two eggs, or more, in a season, by opening a female during the laying-time. If more than one was come down out of the ovary, and advanced to a good size, doubtless then she would that spring lay more than one.

I will endeavour to get a hen, and to examine.

Your supposition that there may be some natural obstruction in singing birds while they are mute, and that when this is removed the song recommences, is new and bold: I wish you could discover some good grounds for this suspicion.*

I was glad you were pleased with my specimen of the *caprimulgus*, or fern-owl; you were, I find, acquainted with the bird before.

When we meet I shall be glad to have some conversation with you concerning the proposal you make of my drawing up an account of the animals in this neighbourhood. Your partiality towards my small abilities persuades you, I fear, that I am able to do more than is in my power: for it is no small undertaking for a man unsupported and alone to begin a natural history from his own autopsia!* Though there is

F. CUCKOO.

endless room for observation in the field of nature, which is bound-less, yet investigation (where a man endeavours to be sure of his facts) can make but slow progress; and all that one could collect in many years would go into a very narrow compass.

Some extracts from your ingenious 'Investigations of the differ-ence between the present temperature of the air in Italy,' &c. have fallen in my way; and gave me great satisfaction: they have removed the objections that always arose in my mind whenever I came to the passages which you quote. Surely the judicious Virgil, when writing a didactic poem for the region of Italy, could never think of describ-ing freezing rivers, unless such severity of weather pretty frequently occurred!*

P. S. Swallows appear amidst snows and frost.

LETTER 6

Selborne, May 21, 1770

Dear Sir,

The severity and turbulence of last month so interrupted the regular process of summer migration, that some of the birds do but just begin to shew themselves, and others are apparently thinner than usual; as the white-throat, the black-cap, the redstart, the fly-catcher. I well remember that after the very severe spring in the year 1739–40 summer birds of passage were very scarce. They come probably hither with a south-east wind, or when it blows between those points; but in that unfavourable year the winds blowed the whole spring and summer through from the opposite quarters. And yet amidst all these disadvantages two swallows, as I mentioned in my last, appeared this year as early as the eleventh of April amidst frost and snow; but they withdrew again for a time.

I am not pleased to find that some people seem so little satisfied with Scopoli's new publication;[1] there is room to expect great things from the hands of that man, who is a good naturalist: and one would think that an history of the birds of so distant and southern a region as Carniola would be new and interesting. I could wish to see that work, and hope to get it sent down. Dr Scopoli is physician to the wretches that work in the quicksilver mines* of that district.

When you talked of keeping a reed-sparrow, and giving it seeds, I could not help wondering; because the reed-sparrow which I mentioned to you (*passer arundinaceus minor Raii*) is a soft-billed bird; and most probably migrates hence before winter; whereas the bird you kept (*passer torquatus Raii*) abides all the year, and is a thick-billed bird. I question whether the latter be much of a songster; but in this matter I want to be better informed. The former has a variety of hurrying notes, and sings all night. Some part of the song of the former, I suspect, is attributed to the latter. We have plenty of the soft-billed sort; which Mr Pennant had entirely left out of his *British Zoology*, till I reminded him of his omission. See *British Zoology* last published, p. 16.[2]

---

[1] This work he calls his *Annus Primus Historico Naturalis* [1769].

[2] See letter 25 to Mr Pennant.

I have somewhat to advance on the different manners in which different birds fly and walk; but as this is a subject that I have not enough considered, and is of such a nature as not to be contained in a small space, I shall say nothing further about it at present.[1]

No doubt the reason why the sex of birds in their first plumage is so difficult to be distinguished is, as you say, 'because they are not to pair and discharge their parental functions till the ensuing spring.' As colours seem to be the chief external sexual distinction in many birds, these colours do not take place till sexual attachments begin to obtain. And the case is the same in quadrupeds; among whom, in their younger days, the sexes differ but little: but, as they advance to maturity, horns and shaggy manes, beards and brawny necks, &c. &c. strongly discriminate the male from the female. We may instance still farther in our own species, where a beard and stronger features are usually characteristic of the male sex: but this sexual diversity does not take place in earlier life; for a beautiful youth shall be so like a beautiful girl that the difference shall not be discernible;

> Quem si puellarum insereres choro,
> Mirè sagaces falleret hospites
> Discrimen obscurum, solutis
> Crinibus, ambiguoque vultu.

> [If you put him into a group of girls, even a wondrously
> wise stranger would fail to notice that he was different,
> disguised by his loosened hair and ambiguous face.]

> [Horace, *Odes*, 2. 5]

LETTER 7

Ringmer, near Lewes, Oct. 8, 1770

Dear Sir,

I am glad to hear that Kuckalm is to furnish you with the birds of Jamaica;* a sight of the hirundines of that hot and distant island would be a great entertainment to me.

---

[1] See letter 42 to Mr Barrington.

The *Anni* of Scopoli are now in my possession; and I have read the *Annus Primus* with satisfaction: for though some parts of this work are exceptionable, and he may advance some mistaken observations; yet the ornithology of so distant a country as Carniola is very curious. Men that undertake only one district are much more likely to advance natural knowledge than those that grasp at more than they can possibly be acquainted with: every kingdom, every province, should have its own monographer.

The reason perhaps why he mentions nothing of Ray's *Ornithology* may be the extreme poverty and distance of his country, into which the works of our great naturalist may have never yet found their way. You have doubts, I know, whether this *Ornithology* is genuine, and really the work of Scopoli: as to myself, I think I discover strong tokens of authenticity; the style corresponds with that of his *Entomology*;* and his characters* of his Ordines and Genera are many of them new, expressive, and masterly. He has ventured to alter some of the Linnaean genera with sufficient shew of reason.

It might perhaps be mere accident that you saw so many swifts and no swallows at Staines; because, in my long observation of those birds, I never could discover the least degree of rivalry or hostility between the species.

Ray remarks that birds of the gallinae order, as cocks and hens, partridges, and pheasants, &c. are pulveratrices, such as dust themselves, using that method of cleansing their feathers, and ridding themselves of their vermin. As far as I can observe, many birds that dust themselves never wash: and I once thought that those birds that wash themselves would never dust; but here I find myself mistaken; for common house-sparrows are great pulveratrices, being frequently seen grovelling and wallowing in dusty roads; and yet they are great washers. Does not the skylark dust?

Query. Might not Mahomet and his followers take one method of purification from these pulveratrices? because I find from travellers of credit, that if a strict mussulman is journeying in a sandy desert where no water is to be found, at stated hours he strips off his clothes, and most scrupulously rubs his body over with sand or dust.

A countryman told me he had found a young fern-owl in the nest of a small bird on the ground; and that it was fed by the little bird. I went to see this extraordinary phenomenon, and found that it was

a young cuckoo hatched in the nest of a titlark: it was become vastly
too big for its nest, appearing

> . . . in tenui re
> Majores pennas nido extendisse . . .
>
> [of slender means, I spread my wings too wide
> for the nest.]   [Horace, *Epistles*, 1.20]

and was very fierce and pugnacious, pursuing my finger, as I teazed it,
for many feet from the nest, and sparring and buffetting with its
wings like a game-cock. The dupe of a dam appeared at a distance,
hovering about with meat in its mouth, and expressing the greatest
solicitude.

In July I saw several cuckoos skimming over a large pond; and
found, after some observation, that they were feeding on the *libellulae*,
or dragon-flies; some of which they caught as they settled on the weeds,
and some as they were on the wing. Notwithstanding what Linnaeus
says, I cannot be induced to believe that they are birds of prey.

This district* affords some birds that are hardly ever heard of at
Selborne. In the first place considerable flocks of cross-beaks (*loxiae
curvirostrae*) have appeared this summer in the pine-groves belonging
to this house; the water-ousel is said to haunt the mouth of the Lewes
river, near Newhaven; and the Cornish chough builds, I know, all
along the chalky cliffs of the Sussex shore.

I was greatly pleased to see little parties of ring-ousels (my newly
discovered migraters) scattered, at intervals, all along the Sussex
downs from Chichester to Lewes. Let them come from whence they
will, it looks very suspicious that they are cantoned* along the coast
in order to pass the channel when severe weather advances. They
visit us again in April, as it should seem, in their return; and are not
to be found in the dead of winter. It is remarkable that they are very
tame, and seem to have no manner of apprehensions of danger from
a person with a gun. There are bustards on the wide downs near
Brighthelmstone.* No doubt you are acquainted with the Sussex
downs: the prospects and rides round Lewes are most lovely!

As I rode along near the coast I kept a very sharp look out in the
lanes and woods, hoping I might, at this time of the year, have discovered some of the summer short-winged birds of passage crowding
towards the coast in order for their departure: but it was very extraordinary that I never saw a redstart, white-throat, black-cap, uncrested

CHOUGH.

wren, fly-catcher, &c. And I remember to have made the same remark in former years, as I usually come to this place annually about this time. The birds most common along the coast at present are the stone-chatters, whinchats, buntings, linnets, some few wheat-ears, titlarks, &c. Swallows and house-martins abound yet, induced to prolong their stay by this soft, still, dry season.

A land tortoise, which has been kept for thirty years in a little walled court belonging to the house where I now am visiting,* retires under ground about the middle of November, and comes forth again about the middle of April. When it first appears in the spring it discovers very little inclination towards food; but in the height of summer grows voracious: and then as the summer declines its appetite declines; so that for the last six weeks in autumn it hardly eats at all. Milky plants, such as lettuces, dandelions, sowthistles, are its favourite dish. In a neighbouring village one was kept till by tradition it was supposed to be an hundred years old. An instance of vast longevity in such a poor reptile!

LETTER 8

Selborne, Dec. 20, 1770

Dear Sir,

There are doubtless many home internal migrations within this kingdom that want to be better understood: witness those vast flocks of hen chaffinches that appear with us in the winter without hardly any cocks among them. Now was there a due proportion of each sex, it should seem very improbable that any one district should produce such numbers of these little birds; and much more when only one half of the species appears: therefore we may conclude that the *fringillae caelebes*, for some good purposes, have a peculiar migration of their own in which the sexes part. Nor should it seem so wonderful that the intercourse* of sexes in this species of birds should be interrupted in winter; since in many animals, and particularly in bucks and does, the sexes herd separately, except at the season when commerce* is necessary for the continuance of the breed. For this matter of the chaffinches see *Fauna Suecica*, p. 85, and *Systema Naturae*, p. 318.* I see every winter vast flights of hen chaffinches, but none of cocks.

Your method of accounting for the periodical motions of the British singing birds, or birds of flight, is a very probable one; since the matter of food is a great regulator of the actions and proceedings of the brute creation: there is but one that can be set in competition with it, and that is love. But I cannot quite acquiesce with you in one circumstance when you advance that, 'when they have thus feasted, they again separate into small parties of five or six, and get the best fare they can within a certain district, having no inducement to go in quest of fresh-turned earth.' Now if you mean that the business of congregating is quite at an end from the conclusion of wheat-sowing to the season of barley and oats,* it is not the case with us; for larks and chaffinches, and particularly linnets, flock and congregate as much in the very dead of winter as when the husbandman is busy with his ploughs and harrows.

Sure there can be no doubt but that woodcocks and fieldfares leave us in the spring, in order to cross the seas, and to retire to some districts more suitable to the purpose of breeding. That the former pair before they retire, and that the hens are forward with egg, I myself, when I was a sportsman, have often experienced. It cannot indeed be denied but that now and then we hear of a woodcock's nest, or young birds, discovered in some part or other of this island: but then they are always mentioned as rarities, and somewhat out of the common course of things:* but as to redwings and fieldfares, no sportsman or naturalist has ever yet, that I could hear, pretended to have found the nest or young of those species in any part of these kingdoms. And I the more admire at this instance as extraordinary, since, to all appearance, the same food in summer as well as in winter might support them here which maintains their congeners, the blackbirds and thrushes, did they chuse to stay the summer through. From hence it appears that it is not food alone which determines some species of birds with regard to their stay or departure. Fieldfares and redwings disappear sooner or later according as the warm weather comes on earlier or later. For I well remember, after that dreadful winter 1739–40, that cold north-east winds continued to blow on through April and May, and that these kinds of birds (what few remained of them) did not depart as usual, but were seen lingering about till the beginning of June.

The best authority that we can have for the nidification of the birds above mentioned in any district, is the testimony of faunists* that

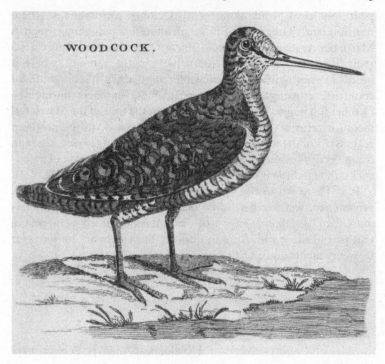

WOODCOCK.

have written professedly the natural history of particular countries. Now, as to the fieldfare, Linnaeus, in his *Fauna Suecica*, says of it that 'maximis in arboribus nidificat [it nests in the tallest trees]': and of the redwing he says, in the same place, that 'nidificat in mediis arbusculis, sive sepibus: ova sex caeruleo-viridia maculis nigris variis [it nests in medium-sized bushes or hedges; it lays six eggs, blue-green in colour with diverse black spots]'. Hence we may be assured that fieldfares and redwings breed in Sweden. Scopoli says, in his *Annus Primus*, of the woodcock, that 'nupta ad nos venit circa aequinoctium vernale [it comes to us, already mated, about the spring equinox]': meaning in Tirol, of which he is a native. And afterwards he adds 'nidificat in paludibus alpinis: ova ponit 3 . . . 5 [it nests in the Alpine swampy woods: it lays from three to five eggs]'. It does not appear from Kramer that woodcocks breed at all in Austria: but he says 'Avis haec septentrionalium provinciarum astivo tempore incola est; ubi plerumque nidificat. Appropinquante hyeme australiores provincias petit: hinc circa plenilunium mensis Octobris plerumque Austriam transmigrant. Tunc rursus circa plenilunium potissimum mensis Martii per Austriam matrimonio juncta ad septentrionales provincias redit [This bird lives in northern parts in the summer, where it breeds in great numbers. When winter approaches, it seeks southern countries, crossing Austria in considerable numbers during the October full moon. Then, if possible about the time of the March full moon, it returns through Austria, already mated, to the northern regions].' For the whole passage (which I have abridged) see *Elenchus*, &c. p. 351. This seems to be a full proof of the migration of wood-cocks; though little is proved concerning the place of breeding.

P.S. There fell in the county of Rutland, in three weeks of this present very wet weather, seven inches and an half of rain, which is more than has fallen in any three weeks for these thirty years past in that part of the world. A mean quantity in that county for one year is twenty inches and an half.*

LETTER 9

Fyfield, near Andover, Feb. 12, 1771

Dear Sir,

You are, I know, no great friend to migration; and the well attested accounts from various parts of the kingdom seem to justify you in your suspicions, that at least many of the swallow kind do not leave us in the winter, but lay themselves up like insects and bats, in a torpid state, and slumber away the more uncomfortable months till the return of the sun and fine weather awakens them.

But then we must not, I think, deny migration in general; because migration certainly does subsist in some places, as my brother in Andalusia* has fully informed me. Of the motions of these birds he has ocular demonstration, for many weeks together, both spring and fall: during which periods myriads of the swallow kind traverse the Straits from north to south, and from south to north, according to the season. And these vast migrations consist not only of hirundines but of bee-birds, hoopoes, *oro pendolos*, or golden thrushes, &c. &c. and also of many of our soft-billed summer birds of passage; and more-over of birds which never leave us, such as all the various sorts of hawks and kites. Old Belon, two hundred years ago, gives a curious account of the incredible armies of hawks and kites which he saw in the spring-time traversing the Thracian Bosphorus from Asia to Europe. Besides the above mentioned, he remarks that the procession is swelled by whole troops of eagles and vultures.*

Now it is no wonder that birds residing in Africa should retreat before the sun as it advances, and retire to milder regions, and especially birds of prey, whose blood being heated with hot animal food,* are more impatient of a sultry climate: but then I cannot help wondering why kites and hawks, and such hardy birds as are known to defy all the severity of England, and even of Sweden and all north Europe, should want to migrate from the south of Europe, and be dissatisfied with the winters of Andalusia.

It does not appear to me that much stress may be laid on the difficulty and hazard that birds must run in their migrations, by reason of vast oceans, cross winds, &c.; because, if we reflect, a bird may travel from England to the equator without launching out and exposing itself to boundless seas, and that by crossing the water at Dover, and again at Gibraltar. And I with the more confidence advance this obvious remark, because my brother has always found that some of his birds, and particularly the swallow kind, are very sparing of their pains in crossing the Mediterranean: for when arrived at Gibraltar they do not

> . . . Rang'd in figure wedge their way,
> . . . And set forth
> Their airy caravan high over seas
> Flying, and over lands with mutual wing
> Easing their flight . . .
>
> Milton [*Paradise Lost*, VII. 426–9]

but scout and hurry along in little detached parties of six or seven in a company; and sweeping low, just over the surface of the land and water, direct their course to the opposite continent at the narrowest passage they can find. They usually slope across the bay to the southwest, and so pass over opposite to Tangier, which, it seems, is the narrowest space.

In former letters we have considered whether it was probable that woodcocks in moon-shiny nights cross the German ocean* from Scandinavia. As a proof that birds of less speed may pass that sea, considerable as it is, I shall relate the following incident, which, though mentioned to have happened so many years ago, was strictly matter of fact:—As some people were shooting in the parish of Trotton, in the county of Sussex, they killed a duck in that dreadful winter 1708–9,* with a silver collar about its neck,[1] on which were engraven the arms of the king of Denmark. This anecdote the rector of Trotton at that time has often told to a near relation of mine; and, to the best of my remembrance, the collar was in the possession of the rector.

At present I do not know any body near the sea-side that will take the trouble to remark at what time of the moon woodcocks first come: if I lived near the sea myself I would soon tell you more of the matter. One thing I used to observe when I was a sportsman, that there were times in which woodcocks were so sluggish and sleepy that they would drop again when flushed just before the spaniels, nay just at the muzzle of a gun that had been fired at them: whether this strange laziness was the effect of a recent fatiguing journey I shall not presume to say.

Nightingales not only never reach Northumberland and Scotland, but also, as I have been always told, Devonshire and Cornwall. In those two last counties we cannot attribute the failure of them to the want of warmth: the defect in the west is rather a presumptive argument that these birds come over to us from the continent at the narrowest passage, and do not stroll so far westward.

Let me hear from your own observation whether skylarks do not dust. I think they do: and if they do, whether they wash also.

The *alauda pratensis** of Ray was the poor dupe that was educating the booby of a cuckoo mentioned in my letter of October last.

Your letter came too late for me to procure a ring-ousel for Mr Tunstal during their autumnal visit; but I will endeavour to get him one when they call on us again in April. I am glad that you and that gentleman saw my Andalusian birds;* I hope they answered your expectation. Royston, or grey crows, are winter birds that come much about the same time with the woodcock: they, like the fieldfare and redwing, have no apparent reason for migration; for as they fare in the winter like their congeners, so might they in all appearance in the summer. Was not Tenant, when a boy, mistaken? did he not find a missel-thrush's nest, and take it for the nest of a fieldfare?

---

[1] I have read a like anecdote of a swan.

The stock-dove, or wood-pigeon, *oenas Raii*, is the last winter bird of passage which appears with us; and is not seen till towards the end of November: about twenty years ago they abounded in the district of Selborne; and strings of them were seen morning and evening that reached a mile or more: but since the beechen woods have been greatly thinned they are much decreased in number. The ring-dove, *palumbus Raii*, stays with us the whole year, and breeds several times through the summer.

Before I received your letter of October last I had just remarked in my journal that the trees were unusually green. This uncommon verdure lasted on late into November; and may be accounted for from a late spring, a cool and moist summer; but more particularly from vast armies of chafers, or tree-beetles, which, in many places, reduced whole woods to a leafless naked state. These trees shot again at Midsummer, and then retained their foliage till very late in the year.

My musical friend, at whose house I am now visiting,* has tried all the owls that are his near neighbours with a pitch-pipe set at concert-pitch, and finds they all hoot in B flat. He will examine the nightingales next spring.

I am, &c. &c.

LETTER 10

Selborne, Aug. 1, 1771

Dear Sir,

From what follows, it will appear that neither owls nor cuckoos keep to one note. A friend remarks that many (most) of his owls hoot in B flat; but that one went almost half a note below A. The pipe he tried their notes by was a common half-crown pitch-pipe, such as masters use for tuning of harpsichords; it was the common London pitch.*

A neighbour of mine, who is said to have a nice ear, remarks that the owls about this village hoot in three different keys, in G flat, or F sharp, in B flat and A flat. He heard two hooting to each other, the one in A flat, and the other in B flat. *Query*: Do these different notes proceed from different species, or only from various individuals? The same person finds upon trial that the note of the cuckoo (of which we have but one species) varies in different individuals; for, about

Selborne wood, he found they were mostly in D: he heard two sing together, the one in D, the other in D sharp, who made a disagreeable concert: he afterwards heard one in D sharp, and about Wolmerforest some in C. As to nightingales, he says that their notes are so short, and their transitions so rapid, that he cannot well ascertain their key. Perhaps in a cage, and in a room, their notes may be more distinguishable. This person has tried to settle the notes of a swift, and of several other small birds, but cannot bring them to any criterion.

As I have often remarked that redwings are some of the first birds that suffer with us in severe weather, it is no wonder at all that they retreat from Scandinavian winters: and much more the *ordo* of grallae, who, all to a bird, forsake the northern parts of Europe at the approach of winter. 'Grallae tanquam conjuratae unanimiter in fugam se conjiciunt; ne earum unicam quidem inter nos habitantem invenire possimus; ut enim aestate in australibus degere nequeunt ob defectum lumbricorum, terramque siccam; ita nec in frigidis ob eandem causam [Waders take flight as an entire flock, as if they have conspired together. It might not be possible to find that even one has made a home with us, for just as in summer they cannot live in southern regions because of the lack of worms and the parched soil, so for the same reason they cannot live in cold regions],' says Ekmarck the Swede, in his ingenious little treatise called *Migrationes Avium*, which by all means you ought to read while your thoughts run on the subject of migration. See *Amoenitates Academicae* vol. 4, p. 565.

Birds may be so circumstanced as to be obliged to migrate in one country and not in another: but the grallae, (which procure their food from marshes and boggy grounds) must in winter forsake the more northerly parts of Europe, or perish for want of food.

I am glad you are making inquiries from Linnaeus concerning the woodcock: it is expected of him that he should be able to account for the motions and manner of life of the animals of his own Fauna.

Faunists, as you observe, are too apt to acquiesce in bare descriptions, and a few synonyms: the reason is plain; because all that may be done at home in a man's study, but the investigation of the life and conversation of animals, is a concern of much more trouble and difficulty, and is not to be attained but by the active and inquisitive, and by those that reside much in the country.

Foreign systematics are, I observe, much too vague in their specific differences; which are almost universally constituted by one or two particular marks, the rest of the description running in general terms. But our countryman, the excellent Mr Ray, is the only describer that conveys some precise idea in every term or word, maintaining his superiority over his followers and imitators in spite of the advantage of fresh discoveries and modern information.

At this distance of years it is not in my power to recollect at what periods woodcocks used to be sluggish or alert when I was a sportsman: but, upon my mentioning this circumstance to a friend, he thinks he has observed them to be remarkably listless against snowy foul weather: if this should be the case, then the inaptitude for flying arises only from an eagerness for food; as sheep are observed to be very intent on grazing against stormy wet evenings.

I am, &c. &c.

LETTER 11

Selborne, Feb. 8, 1772

Dear Sir,

When I ride about in the winter, and see such prodigious flocks of various kinds of birds, I cannot help admiring at these congregations, and wishing that it was in my power to account for those appearances almost peculiar to the season. The two great motives which regulate the proceedings of the brute creation are love and hunger; the former incites animals to perpetuate their kind, the latter induces them to preserve individuals: whether either of these should seem to be the ruling passion in the matter of congregating is to be considered. As to love, that is out of the question at a time of the year when that soft passion is not indulged: besides, during the amorous season, such a jealousy prevails between the male birds that they can hardly bear to be together in the same hedge or field. Most of the singing and elation of spirits of that time seem to me to be the effect of rivalry and emulation: and it is to this spirit of jealousy that I chiefly attribute the equal dispersion of birds in the spring over the face of the country.

Now as to the business of food: as these animals are actuated by instinct to hunt for necessary food, they should not, one would suppose, crowd together in pursuit of sustenance at a time when it is most likely to fail; yet such associations do take place in hard weather chiefly, and thicken as the severity increases. As some kind of self-interest and self-defence is no doubt the motive for the proceeding, may it not arise from the helplessness of their state in such rigorous seasons; as men crowd together, when under great calamities, though they know not why? Perhaps approximation* may dispel some degree of cold; and a crowd may make each individual appear safer from the ravages of birds of prey and other dangers.

If I admire when I see how much congenerous birds love to congregate, I am the more struck when I see incongruous ones in such strict amity. If we do not much wonder to see a flock of rooks usually attended by a train of daws,* yet it is strange that the former should so frequently have a flight of starlings for their satellites. Is it because rooks have a more discerning scent than their attendants, and can lead them to spots more productive of food? Anatomists say that rooks, by reason of two large nerves which run down between the eyes into the upper mandible, have a more delicate feeling in their beaks than other round-billed birds, and can grope for their meat when out of sight. Perhaps then their associates attend them on the motive of interest, as greyhounds wait on the motions of their finders; and as lions are said to do on the yelpings of jackalls. Lapwings and starlings sometimes associate.

LETTER 12

March 9, 1772

Dear Sir,

As a gentleman and myself were walking on the fourth of last November round the sea-banks at Newhaven, near the mouth of the Lewes river, in pursuit of natural knowledge, we were surprised to see three house-swallows gliding very swiftly by us.* That morning was rather chilly, with the wind at north-west; but the tenor of the weather for some time before had been delicate, and the noons remarkably warm. From this incident, and from repeated accounts which I meet with, I am more and more induced to believe that many of the swallow kind do not depart from this island; but lay themselves up in holes and caverns; and do, insect-like and bat-like, come forth at mild times, and then retire again to their *latebrae* [hiding places]. Nor make I the least doubt but that, if I lived at Newhaven, Seaford, Brighthelmstone, or any of those towns near the chalk-cliffs of the Sussex coast, by proper observations, I should see swallows stirring at periods of the winter, when the noons were soft and inviting, and the sun warm and invigorating. And I am the more of this opinion from what I have remarked during some of our late springs, that though some swallows did make their appearance about the usual time, *viz.* the thirteenth or fourteenth of April, yet meeting with an harsh reception, and blustering cold north-east winds, they immediately withdrew, absconding for several days, till the weather gave them better encouragement.

LETTER 13

April 12, 1772

Dear Sir,

While I was in Sussex last autumn my residence was at the village near Lewes, from whence I had formerly the pleasure of writing to you.* On the first of November I remarked that the old tortoise, formerly mentioned, began first to dig the ground in order to the forming its *hybernaculum*, which it had fixed on just beside a great tuft of hepaticas. It scrapes out the ground with its fore-feet, and throws it up over its back with its hind; but the motion of its legs is ridiculously slow, little exceeding the hour-hand of a clock; and suitable to the composure of an animal said to be a whole month in performing one feat of copulation. Nothing can be more assiduous than this creature night and day in scooping the earth, and forcing its great body into the cavity; but, as the noons of that season proved unusually warm and sunny, it was continually interrupted, and called forth by the heat in the middle of the day; and though I continued there till the thirteenth of November, yet the work remained unfinished. Harsher weather, and frosty mornings, would have quickened its operations. No part of its behaviour ever struck me more than the extreme timidity it always expresses with regard to rain; for though it has a shell that would secure it against the wheel of a loaded cart, yet does it discover as much solicitude about rain as a lady dressed in all her best attire, shuffling away on the first sprinklings, and running its head up in a corner. If attended to, it becomes an excellent weather-glass; for as sure as it walks elate, and as it were on tiptoe, feeding with great earnestness in a morning, so sure will it rain before night. It is totally a diurnal animal, and never pretends to stir after it becomes dark. The tortoise, like other reptiles, has an arbitrary stomach as well as lungs; and can refrain from eating as well as breathing for a great part of the year. When first awakened it eats nothing; nor again in the autumn before it retires: through the height of the summer it feeds voraciously, devouring all the food that comes in its way. I was much taken with its sagacity in discerning those that do it kind offices: for, as soon as the good old lady* comes in sight who has waited on it for more than thirty years, it hobbles towards its benefactress with aukward alacrity; but remains inattentive to strangers. Thus not only 'the ox knoweth his owner, and the ass his master's crib,'[1] but the most abject reptile and torpid of beings distinguishes the hand that feeds it, and is touched with the feelings of gratitude!

I am, &c. &c.

P.S. In about three days after I left Sussex the tortoise retired into the ground under the *hepatica*.

LETTER 14

Selborne, March 26, 1773

Dear Sir,

The more I reflect on the στοργή [natural affection] of animals, the more I am astonished at its effects. Nor is the violence of this affection more wonderful than the shortness of its duration. Thus every hen is in her turn the virago of the yard, in proportion to the helplessness of her brood; and will fly in the face of a dog or a sow in defence of those chickens, which in a few weeks she will drive before her with relentless cruelty.

This affection sublimes* the passions, quickens the invention, and sharpens the sagacity of the brute creation. Thus an hen, just become a mother, is no longer that placid bird she used to be, but with feathers standing on end, wings hovering, and clocking note, she runs about like one possessed. Dams will throw themselves in the way of

[1] Isaiah i. 3.

the greatest danger in order to avert it from their progeny. Thus a partridge will tumble along before a sportsman in order to draw away the dogs from her helpless covey. In the time of nidification the most feeble birds will assault the most rapacious. All the hirundines of a village are up in arms at the sight of an hawk, whom they will persecute till he leaves that district. A very exact observer has often remarked that a pair of ravens nesting in the rock of Gibraltar would suffer no vulture or eagle to rest near their station, but would drive them from the hill with an amazing fury: even the blue thrush at the season of breeding would dart out from the clefts of the rocks to chase away the kestril, or the sparrow-hawk. If you stand near the nest of a bird that has young, she will not be induced to betray them by an inadvertent fondness, but will wait about at a distance with meat in her mouth for an hour together.

Should I farther corroborate what I have advanced above by some anecdotes which I probably may have mentioned before in conversation, yet you will, I trust, pardon the repetition for the sake of the illustration.

The flycatcher of the *Zoology* (the *stoparola* of Ray),* builds every year in the vines that grow on the walls of my house. A pair of these little birds had one year inadvertently placed their nest on a naked bough, perhaps in a shady time, not being aware of the inconvenience that followed. But an hot sunny season coming on before the brood was half fledged, the reflection of the wall became insupportable, and must inevitably have destroyed the tender young, had not affection suggested an expedient, and prompted the parent-birds to hover over the nest all the hotter hours, while with wings expanded, and mouths gaping for breath, they screened off the heat from their suffering offspring.

A farther instance I once saw of notable sagacity in a willow-wren, which had built in a bank in my fields. This bird a friend and myself had observed as she sat in her nest; but were particularly careful not to disturb her, though we saw she eyed us with some degree of jealousy.* Some days after as we passed that way we were desirous of remarking how this brood went on; but no nest could be found, till I happened to take up a large bundle of long green moss, as it were, carelessly thrown over the nest in order to dodge the eye of any impertinent intruder.

A still more remarkable mixture of sagacity and instinct occurred to me one day as my people were pulling off the lining of an hotbed, in order to add some fresh dung.* From out of the side of this bed leaped an animal with great agility that made a most grotesque figure; nor was it without great difficulty that it could be taken; when it proved to be a large white-bellied field-mouse with three or four young clinging to her teats by their mouths and feet. It was amazing that the desultory and rapid motions of this dam should not oblige her litter to quit their hold, especially when it appeared that they were so young as to be both naked and blind!

To these instances of tender attachment, many more of which might be daily discovered by those that are studious of nature, may be opposed that rage of affection, that monstrous perversion of the στοργή, which induces some females of the brute creation to devour their young because their owners have handled them too freely, or removed them from place to place! Swine, and sometimes the more gentle race of dogs and cats, are guilty of this horrid and preposterous* murder. When I hear now and then of an abandoned mother that destroys her offspring, I am not so much amazed; since reason perverted, and the bad passions let loose, are capable of any enormity: but why the parental feelings of brutes, that usually flow in one most uniform tenor, should sometimes be so extravagantly diverted, I leave to abler philosophers than myself to determine.

I am, &c.

LETTER 15

Selborne, July 8, 1773

Dear Sir,

Some young men went down lately to a pond on the verge of Wolmer-forest to hunt flappers, or young wild-ducks, many of which they caught, and, among the rest, some very minute yet well-fledged wild-fowls alive, which upon examination I found to be teals. I did not know till then that teals ever bred in the south of England, and was much pleased with the discovery: this I look upon as a great stroke* in natural history.

We have had, ever since I can remember, a pair of white owls that constantly breed under the eaves of this church. As I have paid good attention to the manner of life of these birds during their season of breeding, which lasts the summer through, the following remarks may not perhaps be unacceptable:—About an hour before sunset (for then the mice begin to run) they sally forth in quest of prey, and hunt all round the hedges of meadows and small enclosures for them, which seem to be their only food. In this irregular country we can stand on an eminence and see them beat the fields over like a setting-dog, and often drop down in the grass or corn. I have minuted these birds with my watch for an hour together, and have found that they return to their nest, the one or the other of them, about once in five minutes; reflecting at the same time on the adroitness that every animal is possessed of as far as regards the well being of itself and offspring. But a piece of address,* which they shew when they return loaded, should not, I think, be passed over in silence.—As they take their prey with their claws, so they carry it in their claws to their nest: but, as the feet are necessary in their ascent under the tiles, they constantly perch first on the roof of the chancel, and shift the mouse from their claws to their bill, that the feet may be at liberty to take hold of the plate on the wall as they are rising under the eaves.

White owls seem not (but in this I am not positive) to hoot at all: all that clamorous hooting appears to me to come from the wood kinds. The white owl does indeed snore and hiss in a tremendous manner; and these menaces well answer the intention of intimidating: for I have known a whole village up in arms on such an occasion, imagining the church-yard to be full of goblins and spectres. White owls also often scream horribly as they fly along; from this screaming probably arose the common people's imaginary species of screech-owl, which they superstitiously think attends the windows of dying persons. The plumage of the remiges* of the wings of every species of owl that I have yet examined is remarkably soft and pliant. Perhaps it may be necessary that the wings of these birds should not make much resistance or rushing, that they may be enabled to steal through the air unheard upon a nimble and watchful quarry.

While I am talking of owls, it may not be improper to mention what I was told by a gentleman of the county of Wilts. As they were grubbing a vast hollow pollard-ash that had been the mansion of owls for centuries, he discovered at the bottom a mass of matter that at first he could not account for. After some examination, he found that it was a congeries* of the bones of mice (and perhaps of birds and bats) that had been heaping together for ages, being cast up in pellets out of the crops of many generations of inhabitants. For owls cast up the bones, fur, and feathers, of what they devour, after the manner of hawks. He believes, he told me, that there were bushels* of this kind of substance.

When brown owls hoot their throats swell as big as an hen's egg. I have known an owl of this species live a full year without any water. Perhaps the case may be the same with all birds of prey. When owls fly they stretch out their legs behind them as a balance to their large heavy heads: for as most nocturnal birds have large eyes and ears they must have large heads to contain them. Large eyes I presume are necessary to collect every ray of light, and large concave ears to command the smallest degree of sound or noise.

I am, &c.

It will be proper to premise here that the sixteenth, eighteenth, twentieth, and twenty-first letters have been published already in the *Philosophical Transactions*:* but as nicer observation has furnished several corrections and additions, it is hoped that the republication of them will not give offence; especially as these sheets would be very imperfect without them, and as they will be new to many readers who had no opportunity of seeing them when they made their first appearance.

The hirundines are a most inoffensive, harmless, entertaining, social, and useful tribe of birds: they touch no fruit in our gardens; delight, all except one species, in attaching themselves to our houses; amuse us with their migrations, songs, and marvellous agility; and clear our outlets* from the annoyances of gnats and other troublesome insects. Some districts in the south seas, near Guiaquil,[1] are desolated, it seems, by the infinite swarms of venomous mosquitoes, which fill the air, and render those coasts insupportable. It would be worth inquiring whether any species of hirundines is found in those regions. Whoever contemplates the myriads of insects that sport in the sun-beams of a summer evening in this country, will soon be convinced to what a degree our atmosphere would be choaked with them was it not for the friendly interposition of the swallow tribe.

Many species of birds have their peculiar lice; but the hirundines

---

[1] See Ulloa's travels.*

alone seem to be annoyed with dipterous* insects, which infest every species, and are so large, in proportion to themselves, that they must be extremely irksome and injurious to them. These are the *hippoboscae hirundinis*, with narrow subulated* wings, abounding in every nest; and are hatched by the warmth of the bird's own body during incubation, and crawl about under its feathers.

A species of them is familiar to horsemen in the south of England under the name of forest-fly; and to some of side-fly, from its running sideways like a crab. It creeps under the tails, and about the groins, of horses, which, at their first coming out of the north, are rendered half frantic by the tickling sensation; while our own breed little regards them.

The curious Reaumur discovered the large eggs, or rather pupae* of these flies as big as the flies themselves, which he hatched in his own bosom. Any person that will take the trouble to examine the old nests of either species of swallows may find in them the black shining cases or skins of the pupae of these insects: but for other particulars, too long for this place, we refer the reader to *l'Histoire d'Insectes* of that admirable entomologist. Tom. iv, pl. II.*

LETTER 16

Selborne, Nov. 20. 1773

Dear Sir,

In obedience to your injunctions I sit down to give you some account of the house-martin, or martlet; and, if my monography* of this little domestic and familiar bird should happen to meet with your approbation, I may probably soon extend my inquiries to the rest of the British hirundines—the swallow, the swift, and the bank-martin.

A few house-martins begin to appear about the sixteenth of April; usually some few days later than the swallow. For some time after they appear the hirundines in general pay no attention to the business of nidification, but play and sport about, either to recruit from the fatigue of their journey, if they do migrate at all, or else that their blood may recover its true tone and texture after it has been so long benumbed by the severities of winter. About the middle of May, if the weather be fine, the martin begins to think in earnest of providing a mansion for its family. The crust or shell of this nest seems to be formed of such dirt or loam as comes most readily to hand, and is tempered and wrought together with little bits of broken straws to render it tough and tenacious. As this bird often builds against a perpendicular wall without any projecting ledge under, it requires its utmost efforts to get the first foundation firmly fixed, so that it may safely carry the superstructure. On this occasion the bird not only clings with its claws, but partly supports itself by strongly inclining its tail against the wall, making that a fulcrum; and thus steadied it works and plasters the materials into the face of the brick or stone. But then, that this work may not, while it is soft and green, pull itself down by its own weight, the provident architect has prudence and forbearance enough not to advance her work too fast; but by building only in the morning, and by dedicating the rest of the day to food and amusement, gives it sufficient time to dry and harden. About half an inch seems to be a sufficient layer for a day. Thus careful workmen when they build mud-walls (informed at first perhaps by this little bird) raise but a moderate layer at a time, and then desist; lest the work should become top-heavy, and so be ruined by its own weight. By this method in about ten or twelve days is formed an hemispheric nest with a small aperture towards the top, strong, compact, and warm; and perfectly fitted for all the purposes for which it was intended. But then nothing is more common than for the house-sparrow, as soon as the shell is finished, to seize on it as its own, to eject the owner, and to line it after its own manner.

After so much labour is bestowed in erecting a mansion, as Nature seldom works in vain, martins will breed on for several years together in the same nest, where it happens to be well sheltered and secure from the injuries of weather. The shell or crust of the nest is a sort of rustic-work full of knobs and protuberances on the outside: nor is the inside of those that I have examined smoothed with any exactness at all; but is rendered soft and warm, and fit for incubation, by a lining of small straws, grasses, and feathers; and sometimes by a bed of moss interwoven with wool. In this nest they tread,* or engender, frequently during the time of building; and the hen lays from three to five white eggs.

At first when the young are hatched, and are in a naked and helpless condition, the parent birds, with tender assiduity, carry out what comes away from their young. Was it not for this affectionate cleanliness the nestlings would soon be burnt up, and destroyed in so deep and hollow a nest, by their own caustic excrement. In the quadruped creation the same neat precaution is made use of; particularly among dogs and cats, where the dams lick away what proceeds from their young. But in birds there seems to be a particular provision, that the dung of nestlings is enveloped in a tough kind of jelly, and therefore is the easier conveyed off without soiling or daubing. Yet, as nature is cleanly in all her ways, the young perform this office for themselves in a little time by thrusting their tails out at the aperture of their nest. As the young of small birds presently arrive at their ἡλικία, or full growth, they soon become impatient of confinement, and sit all day with their heads out at the orifice, where the dams, by clinging to the nest, supply them with food from morning to night. For a time the young are fed on the wing by their parents; but the feat is done by so quick and almost imperceptible a slight, that a person must have attended very exactly to their motions before he would be able to perceive it. As soon as the young are able to shift for themselves, the dams immediately turn their thoughts to the business of a second brood: while the first flight, shaken off and rejected by their nurses, congregate in great flocks, and are the birds that are seen clustering and hovering on sunny mornings and evenings round towers and steeples, and on the roofs of churches and houses. These congregatings usually begin to take place about the first week in August; and therefore we may conclude that by that time the first flight is pretty well over. The young of this species do not quit their abodes all together; but the more forward birds get abroad some days before the rest. These approaching the eaves of buildings, and playing about before them, make people think that several old ones attend one nest. They are often capricious in fixing on a nesting-place, beginning many edifices, and leaving them unfinished; but when once a nest is completed in a sheltered place, it serves for several seasons. Those which breed in a ready finished house get the start in hatching of those that build new by ten days or a fortnight. These industrious artificers are at their labours in the long days before four in the morning: when they fix their materials they plaster them on with their chins, moving their heads with a quick vibratory motion. They dip and wash as they fly sometimes in very hot weather, but not so frequently as swallows. It has been observed that martins usually build to a north-east or north-west aspect, that the heat of the sun may not

crack and destroy their nests: but instances are also remembered
where they bred for many years in vast abundance in an hot stifled
inn-yard, against a wall facing to the south.

Birds in general are wise in their choice of situation: but in this
neighbourhood every summer is seen a strong proof to the contrary
at an house without eaves in an exposed district, where some martins
build year by year in the corners of the windows. But, as the corners
of these windows (which face to the south-east and south-west) are
too shallow, the nests are washed down every hard rain; and yet these
birds drudge on to no purpose from summer to summer, without
changing their aspect or house. It is a piteous sight to see them
labouring when half their nest is washed away and bringing dirt . . .
'generis lapsi sarcire ruinas [to repair the ruin of their fallen state.
Virgil, *Georgics* 4]'. Thus is instinct a most wonderful unequal faculty; in some instances so much above reason, in other respects so far
below it! Martins love to frequent towns, especially if there are great
lakes and rivers at hand; nay they even affect the close air of London.*
And I have not only seen them nesting in the Borough,* but even in
the Strand and Fleet-street; but then it was obvious from the dingi-
ness of their aspect that their feathers partook of the filth of that sooty
atmosphere. Martins are by far the least agile of the four species; their
wings and tails are short, and therefore they are not capable of such
surprising turns and quick and glancing evolutions as the swallow.
Accordingly they make use of a placid easy motion in a middle region
of the air, seldom mounting to any great height, and never sweeping
long together over the surface of the ground or water. They do not
wander far for food, but affect sheltered districts, over some lake, or
under some hanging wood, or in some hollow vale, especially in
windy weather. They breed the latest of all the swallow kind: in 1772
they had nestlings on to October the twenty-first, and are never without unfledged young as late as Michaelmas.*

As the summer declines the congregating flocks increase in numbers daily by the constant accession of the second broods; till at last
they swarm in myriads upon myriads round the villages on the
Thames, darkening the face of the sky as they frequent the aits of that
river, where they roost. They retire, the bulk of them I mean, in vast
flocks together about the beginning of October: but have appeared
of late years in a considerable flight in this neighbourhood, for one
day or two, as late as November the third and sixth, after they were

supposed to have been gone for more than a fortnight. They therefore withdraw with us the latest of any species. Unless these birds are very short-lived indeed, or unless they do not return to the district where they are bred, they must undergo vast devastations some how, and some where; for the birds that return yearly bear no manner of proportion to the birds that retire.

House-martins are distinguished from their congeners by having their legs covered with soft downy feathers down to their toes. They are no songsters; but twitter in a pretty inward soft manner in their nests. During the time of breeding they are often greatly molested with fleas.

I am, &c.

LETTER 17

Ringmer, near Lewes, Dec. 9, 1773
Dear Sir,

I received your last favour just as I was setting out for this place; and am pleased to find that my monography* met with your approbation. My remarks are the result of many years observation; and are, I trust, true in the whole: though I do not pretend to say that they are perfectly void of mistake, or that a more nice observer might not make many additions, since subjects of this kind are inexhaustible.

If you think my letter worthy the notice of your respectable society, you are at liberty to lay it before them;* and they will consider it, I hope, as it was intended, as an humble attempt to promote a more minute inquiry into natural history; into the life and conversation of animals. Perhaps hereafter I may be induced to take the house-swallow under consideration; and from that proceed to the rest of the British hirundines.

Though I have now travelled the Sussex-downs upwards of thirty years, yet I still investigate that chain of majestic mountains* with fresh admiration year by year; and think I see new beauties every time I traverse it. This range, which runs from Chichester eastward as far as East-Bourn, is about sixty miles in length, and is called the South Downs, properly speaking, only round Lewes. As you pass along you command a noble view of the wild, or weald,* on one hand, and the broad downs and sea on the other. Mr Ray used to visit a family[1] just at the foot of these hills, and was so ravished with the prospect from Plumpton-plain, near Lewes, that he mentions those scapes in his *Wisdom of God in the Works of the Creation* with the utmost satisfaction, and thinks them equal to any thing he had seen in the finest parts of Europe.

For my own part, I think there is somewhat peculiarly sweet and amusing in the shapely figured aspect of chalk-hills in preference to those of stone, which are rugged, broken, abrupt, and shapeless.

Perhaps I may be singular in my opinion, and not so happy as to convey to you the same idea; but I never contemplate these mountains without thinking I perceive somewhat analogous to growth in their gentle swellings and smooth fungus-like protuberances, their fluted sides, and regular hollows and slopes, that carry at once the air of vegetative dilatation and expansion - - - - - -

- - - Or was there ever a time when these immense masses of calcarious matter were thrown into fermentation by some adventitious moisture; were raised and leavened into such shapes by some plastic power;* and so made to swell and heave their broad backs into the sky so much above the less animated clay of the wild below?

By what I can guess from the admeasurements of the hills that have been taken round my house, I should suppose that these hills surmount the wild at an average at about the rate of five hundred feet.

One thing is very remarkable as to the sheep: from the westward till you get to the river Adur all the flocks have horns, and smooth white faces, and white legs; and a hornless sheep is rarely to be seen: but as soon as you pass that river eastward, and mount Beeding-hill, all the flocks at once become hornless, or, as they call them, poll-sheep; and have moreover black faces with a white tuft of wool on their foreheads, and speckled and spotted legs: so that you would think that the flocks of Laban were pasturing on one side of the stream, and the variegated breed of his son-in-law Jacob were cantoned along on the other.* And this diversity holds good respectively on each side from the valley of Bramber and Beeding to the eastward, and westward all the whole length of the downs. If you talk with the shepherds on this subject, they tell you that the case has been so

---

[1] Mr Courthope of Danny.

from time immemorial; and smile at your simplicity if you ask them whether the situation of these two different breeds might not be reversed? However, an intelligent friend of mine near Chichester* is determined to try the experiment; and has this autumn, at the hazard of being laughed at, introduced a parcel of black-faced hornless rams among his horned western ewes. The black-faced poll-sheep have the shortest legs and the finest wool.

As I had hardly ever before travelled these downs at so late a season of the year, I was determined to keep as sharp a look-out as possible so near the southern coast, with respect to the summer short-winged birds of passage. We make great inquiries concerning the withdrawing of the swallow kind, without examining enough into the causes why this tribe is never to be seen in winter: for, *entre nous*, the disappearing of the latter is more marvellous than that of the former, and much more unaccountable. The hirundines, if they please, are certainly capable of migration; and yet no doubt are often found in a torpid state: but redstarts, nightingales, white-throats, black-caps, &c. &c. are very ill provided for long flights; have never been once found, as I ever heard of, in a torpid state, and yet can never be supposed, in such troops, from year to year to dodge and elude the eyes of the curious and inquisitive, which from day to day discern the other small birds that are known to abide our winters. But, notwithstanding all my care, I saw nothing like a summer bird of passage: and, what is more strange, not one wheat-ear, though they abound so in the autumn as to be a considerable perquisite to the shepherds that take them; and though many are to be seen to my knowledge all the winter through in many parts of the south of England. The most intelligent shepherds tell me that some few of these birds appear on the downs in March, and then withdraw to breed probably in warrens and stone-quarries: now and then a nest is plowed up in a fallow on the downs under a furrow, but it is thought a rarity. At the time of wheat-harvest they begin to be taken in great numbers; are sent for sale in vast quantities to Brighthelmstone and Tunbridge; and appear at the tables of all the gentry that entertain with any degree of elegance. About Michaelmas they retire and are seen no more till March. Though these birds are, when in season, in great plenty on the south downs round Lewes, yet at East-Bourn, which is the eastern extremity of those downs, they abound much more. One thing is very remarkable—that though in the height of the

RING OUZEL.

season so many hundreds of dozens are taken, yet they never are seen to flock; and it is a rare thing to see more than three or four at a time: so that there must be a perpetual flitting and constant progressive succession. It does not appear that any wheat-ears are taken to the westward of Houghton-bridge, which stands on the river Arun.

I did not fail to look particularly after my new migration of ring-ousels; and to take notice whether they continued on the downs to this season of the year; as I had formerly remarked them in the month of October all the way from Chichester to Lewes where-ever there were any shrubs and covert: but not one bird of this sort came within my observation. I only saw a few larks and whin-chats, some rooks, and several kites and buzzards.

About Midsummer a flight of cross-bills comes to the pine-groves about this house, but never makes any long stay.

The old tortoise, that I have mentioned in a former letter, still continues in this garden; and retired under ground about the twentieth of November, and came out again for one day on the thirtieth: it lies now buried in a wet swampy border under a wall facing to the south, and is enveloped at present in mud and mire!

Here is a large rookery round this house, the inhabitants of which seem to get their livelihood very easily; for they spend the greatest part of the day on their nest-trees when the weather is mild. These rooks retire every evening all the winter from this rookery, where they only call by the way, as they are going to roost in deep woods: at the dawn of day they always revisit their nest-trees, and are preceded a few minutes by a flight of daws, that act, as it were, as their harbingers.

I am, &c.

LETTER 18

Selborne, Jan. 29, 1774

Dear Sir,

The house-swallow, or chimney-swallow, is undoubtedly the first comer of all the British hirundines; and appears in general on or about the thirteenth of April, as I have remarked from many years observation. Not but now and then a straggler is seen much earlier: and, in particular, when I was a boy I observed a swallow for a whole day together on a sunny warm Shrove Tuesday; which day could not fall out later than the middle of March, and often happened early in February.*

It is worth remarking that these birds are seen first about lakes and mill-ponds; and it is also very particular, that if these early visiters happen to find frost and snow, as was the case of the two dreadful springs of 1770 and 1771, they immediately withdraw for a time. A circumstance this much more in favour of hiding than migration; since it is much more probable that a bird should retire to its *hybernaculum* just at hand, than return for a week or two only to warmer latitudes.

The swallow, though called the chimney-swallow, by no means

builds altogether in chimneys, but often within barns and out-houses against the rafters; and so she did in Virgil's time:

> . . . Antè
> Garrula quam tignis nidos suspendat hirundo.
>
> [Before the twittering swallow hangs its nest from
> the rafters.]   [Virgil, *Georgics* 4]

In Sweden she builds in barns, and is called *ladu swala*, the barn-swallow. Besides, in the warmer parts of Europe there are no chimnies to houses, except they are English-built: in these countries she constructs her nest in porches, and gate-ways, and galleries, and open halls.

Here and there a bird may affect some odd, peculiar place; as we have known a swallow build down the shaft of an old well, through which chalk had been formerly drawn up for the purpose of manure: but in general with us this hirundo breeds in chimnies; and loves to haunt those stacks where there is a constant fire, no doubt for the sake of warmth. Not that it can subsist in the immediate shaft where there is a fire; but prefers one adjoining to that of the kitchen, and disregards the perpetual smoke of that funnel, as I have often observed with some degree of wonder.

Five or six or more feet down the chimney does this little bird begin to form her nest about the middle of May, which consists, like that of the house-martin, of a crust or shell composed of dirt or mud, mixed with short pieces of straw to render it tough and permanent; with this difference, that whereas the shell of the martin is nearly hemispheric, that of the swallow is open at the top, and like half a deep dish: this nest is lined with fine grasses, and feathers which are often collected as they float in the air.

Wonderful is the address* which this adroit bird shews all day long in ascending and descending with security through so narrow a pass. When hovering over the mouth of the funnel, the vibrations of her wings acting on the confined air occasion a rumbling like thunder. It is not improbable that the dam submits to this inconvenient situation so low in the shaft, in order to secure her broods from rapacious birds, and particularly from owls, which frequently fall down chimnies, perhaps in attempting to get at these nestlings.

The swallow lays from four to six white eggs, dotted with red specks; and brings out her first brood about the last week in June, or the first week in July. The progressive method by which the young are introduced into life is very amusing: first, they emerge from the shaft with difficulty enough, and often fall down into the rooms below: for a day or so they are fed on the chimney-top, and then are conducted to the dead leafless bough of some tree, where, sitting in a row, they are attended with great assiduity, and may then be called perchers. In a day or two more they become flyers, but are still unable to take their own food; therefore they play about near the place where the dams are hawking for flies; and, when a mouthful is collected, at a certain signal given, the dam and the nestling advance, rising towards each other, and meeting at an angle; the young one all the while uttering such a little quick note of gratitude and complacency,* that a person must have paid very little regard to the wonders of Nature that has not often remarked this feat.

The dam betakes herself immediately to the business of a second brood as soon as she is disengaged from her first; which at once associates with the first broods of house-martins; and with them congregates, clustering on sunny roofs, towers, and trees. This hirundo brings out her second brood towards the middle and end of August.

All the summer long is the swallow a most instructive pattern of unwearied industry and affection; for, from morning to night, while there is a family to be supported, she spends the whole day in skimming close to the ground, and exerting the most sudden turns and quick evolutions. Avenues, and long walks under hedges, and pasture-fields, and mown meadows where cattle graze, are her delight, especially if there are trees interspersed; because in such spots insects most abound. When a fly is taken a smart snap from her bill is heard, resembling the noise at the shutting of a watch-case; but the motion of the mandibles are too quick for the eye.

The swallow, probably the male bird, is the *excubitor* [sentinel] to house-martins, and other little birds, announcing the approach of birds of prey. For as soon as an hawk appears, with a shrill alarming note he calls all the swallows and martins about him; who pursue in a body, and buffet and strike their enemy till they have driven him from the village, darting down from above on his back, and rising in a perpendicular line in perfect security. This bird also will sound the alarm, and strike at cats when they climb on the roofs of houses, or otherwise approach the nests. Each species of hirundo drinks as it flies along, sipping the surface of the water; but the swallow alone, in general, washes on the wing, by dropping into a pool for many times together: in very hot weather house-martins and bank-martins dip and wash a little.

The swallow is a delicate songster, and in soft sunny weather sings both perching and flying; on trees in a kind of concert, and on chimney tops: is also a bold flyer, ranging to distant downs and commons even in windy weather, which the other species seem much to dislike; nay, even frequenting exposed sea-port towns, and making little excursions over the salt water. Horsemen on wide downs are often closely attended by a little party of swallows for miles together, which plays before and behind them, sweeping around, and collecting all the sculking insects that are roused by the trampling of the horses feet: when the wind blows hard, without this expedient, they are often forced to settle to pick up their lurking prey.

This species feeds much on little coleoptera, as well as on gnats and flies; and often settles on dug ground, or paths, for gravels to grind and digest its food. Before they depart, for some weeks, to a bird, they forsake houses and chimnies, and roost in trees; and usually withdraw about the beginning of October; though some few stragglers may appear on at times till the first week in November.

Some few pairs haunt the new and open streets of London next the fields, but do not enter, like the house-martin, the close and crowded parts of the city.

Both male and female are distinguished from their congeners by the length and forkedness of their tails. They are undoubtedly the most nimble of all the species: and when the male pursues the female in amorous chase, they then go beyond their usual speed, and exert a rapidity almost too quick for the eye to follow.

After this circumstantial detail of the life and discerning στοργή [natural affection] of the swallow, I shall add, for your farther amusement, an anecdote or two not much in favour of her sagacity:—

A certain swallow built for two years together on the handles of a pair of garden-shears, that were stuck up against the boards in an out-house, and therefore must have her nest spoiled whenever that implement was wanted: and, what is stranger still, another bird of the same species built its nest on the wings and body of an owl that happened by accident to hang dead and dry from the rafter of a barn. This owl, with the nest on its wings, and with eggs in the nest, was brought as a curiosity worthy the most elegant private museum in Great-Britain. The owner, struck with the oddity of the sight, furnished the bringer with a large shell, or conch, desiring him to fix it just where the owl hung: the person did as he was ordered, and the following year a pair, probably the same pair, built their nest in the conch, and laid their eggs.

The owl and the conch make a strange grotesque appearance, and are not the least curious specimens in that wonderful collection of art and nature.[1]

Thus is instinct in animals, taken the least out of its way, an undistinguishing, limited faculty; and blind to every circumstance that does not immediately respect self-preservation, or lead at once to the propagation or support of their species.

I am,

With all respect, &c. &c.

LETTER 19

Selborne, Feb. 14, 1774

Dear Sir,

I received your favour of the eighth, and am pleased to find that you read my little history of the swallow with your usual candour: nor was I the less pleased to find that you made objections where you saw reason.

As to the quotations, it is difficult to say precisely which species of *hirundo* Virgil might intend in the lines in question, since the ancients did not attend to specific differences like modern naturalists: yet somewhat may be gathered, enough to incline me to suppose that in the two passages quoted the poet had his eye on the swallow.

In the first place the epithet *garrula* suits the swallow well, who is a great songster; and not the martin, which is rather a mute bird; and when it sings is so inward as scarce to be heard. Besides, if *tignum* in that place signifies a rafter rather than a beam, as it seems to me to do, then I think it must be the swallow that is alluded to, and not the martin; since the former does frequently build within the roof against the rafters; while the latter always, as far as I have been able to observe, builds without the roof against eaves and cornices.

[1] Sir Ashton Lever's Musaeum.*

As to the simile, too much stress must not be laid on it: yet the epithet *nigra* speaks plainly in favour of the swallow, whose back and wings are very black; while the rump of the martin is milk-white, its back and wings blue, and all its under part white as snow. Nor can the clumsy motions (comparatively clumsy) of the martin well represent the sudden and artful evolutions and quick turns which Juturna gave to her brother's chariot, so as to elude the eager pursuit of the enraged Aeneas.* The verb *sonat* also seems to imply a bird that is somewhat loquacious.[1]

We have had a very wet autumn and winter, so as to raise the springs to a pitch beyond any thing since 1764; which was a remarkable year for floods and high waters. The land-springs, which we call lavants, break out much on the downs of Sussex, Hampshire, and Wiltshire. The country people say when the lavants rise corn will always be dear; meaning that when the earth is so glutted with water as to send forth springs on the downs and uplands, that the corn-vales must be drowned: and so it has proved for these ten or eleven years past. For land-springs have never obtained more since the memory of man than during that period; nor has there been known a greater scarcity of all sorts of grain, considering the great improvements of modern husbandry. Such a run of wet seasons a century or two ago would, I am persuaded, have occasioned a famine. Therefore pamphlets and newspaper letters, that talk of combinations, tend to inflame and mislead;* since we must not expect plenty till Providence sends us more favourable seasons.

The wheat of last year, all round this district, and in the county of Rutland, and elsewhere, yields remarkably bad: and our wheat on the ground, by the continual late sudden vicissitudes from fierce frost to pouring rains, looks poorly; and the turnips rot very fast.

I am, &c.

[1] Nigra velut magnas domini cum divitis aedes
Pervolat, et pennis alta atria lustrat hirundo,
Pabula parva legens, nidisque loquacibus escas:
Et nunc porticibus vacuis, nunc humida circum
Stagna sonat . . . .

[As when a blue-black swallow sweeps through the great mansion of a rich lord and flies amidst its lofty rooms, seizing scraps of sustenance for her noisy nestlings, so can we hear her twittering calls echo now through the spacious entranceways, now over the canals of the water-gardens. Virgil, *Aeneid*, 12.]

LETTER 20

Selborne, Feb. 26, 1774

Dear Sir,

The sand-martin, or bank-martin, is by much the least of any of the British hirundines; and, as far as we have ever seen, the smallest known *hirundo*: though Brisson asserts that there is one much smaller, and that is the *hirundo esculenta*.

But it is much to be regretted that it is scarce possible for any observer to be so full and exact as he could wish in reciting the circumstances attending the life and conversation of this little bird, since it is *fera naturâ* [a wild creature], at least in this part of the kingdom, disclaiming all domestic attachments, and haunting wild heaths and commons where there are large lakes: while the other species, especially the swallow and house-martin, are remarkably gentle and domesticated, and never seem to think themselves safe but under the protection of man.

Here are in this parish, in the sand-pits and banks of the lakes of Wolmer-forest, several colonies of these birds; and yet they are never seen in the village; nor do they at all frequent the cottages that are scattered about in that wild district. The only instance I ever remember where this species haunts any building is at the town of Bishop's Waltham, in this county, where many sand-martins nestle and breed in the scaffold-holes of the back-wall of William of Wykeham's stables: but then this wall stands in a very sequestered and retired enclosure, and faces upon a large and beautiful lake. And indeed this species seems so to delight in large waters, that no instance occurs of their abounding, but near vast pools or rivers: and in particular it has been remarked that they swarm in the banks of the Thames in some places below London-bridge.

It is curious to observe with what different degrees of architectonic skill Providence has endowed birds of the same genus, and so nearly correspondent in their general mode of life! for while the swallow and the house-martin discover the greatest address in raising and securely fixing crusts or shells of loam as cunabula* for their young, the bank-martin terebrates* a round and regular hole in the sand or earth, which is serpentine, horizontal, and about two feet deep. At the inner end of this burrow does this bird deposit, in a good degree

of safety, her rude nest, consisting of fine grasses and feathers, usually goose-feathers, very inartificially* laid together.

Perseverance will accomplish any thing: though at first one would
be disinclined to believe that this weak bird, with her soft and tender
bill and claws, should ever be able to bore the stubborn sand-bank
without entirely disabling herself: yet with these feeble instruments
have I seen a pair of them make great dispatch: and could remark how
much they had scooped that day by the fresh sand which ran down
the bank, and was of a different colour from that which lay loose and
bleached in the sun.

In what space of time these little artists are able to mine and finish
these cavities I have never been able to discover, for reasons given
above; but it would be a matter worthy of observation, where it falls
in the way of any naturalist to make his remarks. This I have often
taken notice of, that several holes of different depths are left unfinished at the end of summer. To imagine that these beginnings were
intentionally made in order to be in the greater forwardness for next
spring, is allowing perhaps too much foresight and *rerum prudentia*
[circumspection] to a simple bird. May not the cause of these *latebrae*
[hiding-places] being left unfinished arise from their meeting in those
places with strata too harsh, hard, and solid, for their purpose, which
they relinquish, and go to a fresh spot that works more freely? Or may
they not in other places fall in with a soil as much too loose and
mouldering, liable to flounder, and threatening to overwhelm them
and their labours?

One thing is remarkable—that, after some years, the old holes are
forsaken and new ones bored; perhaps because the old habitations
grow foul and fetid from long use, or because they may so abound
with fleas as to become untenantable. This species of swallow moreover is strangely annoyed with fleas: and we have seen fleas, bed-fleas
(*pulex irritans*), swarming at the mouths of these holes, like bees on
the stools* of their hives.

The following circumstance should by no means be omitted—that
these birds do *not* make use of their caverns by way of *hybernacula*, as
might be expected; since banks so perforated have been dug out with
care in the winter, when nothing was found but empty nests.

The sand-martin arrives much about the same time with the
swallow, and lays, as she does, from four to six white eggs. But as
this species is cryptogame,* carrying on the business of nidification,

incubation, and the support of its young in the dark, it would not be so easy to ascertain the time of breeding, were it not for the coming forth of the broods, which appear much about the time, or rather somewhat earlier than those of the swallow. The nestlings are supported in common like those of their congeners, with gnats and other small insects; and sometimes they are fed with *libellulae* (dragon-flies) almost as long as themselves. In the last week in June we have seen a row of these sitting on a rail near a great pool as perchers; and so young and helpless, as easily to be taken by hand: but whether the dams ever feed them on the wing, as swallows and house-martins do, we have never yet been able to determine; nor do we know whether they pursue and attack birds of prey.

When they happen to breed near hedges and enclosures, they are dispossessed of their breeding holes by the house-sparrow, which is on the same account a fell adversary to house-martins.

These hirundines are no songsters, but rather mute, making only a little harsh noise when a person approaches their nests. They seem not to be of a sociable turn, never with us congregating with their congeners in the autumn. Undoubtedly they breed a second time, like the house-martin and swallow; and withdraw about Michaelmas.

Though in some particular districts they may happen to abound, yet in the whole, in the south of England at least, is this much the rarest species. For there are few towns or large villages but what abound with house-martins; few churches, towers, or steeples, but what are haunted by some swifts; scarce a hamlet or single cottage-chimney that has not its swallow; while the bank-martins, scattered here and there, live a sequestered life among some abrupt sand-hills, and in the banks of some few rivers.

These birds have a peculiar manner of flying; flitting about with odd jerks, and vacillations, not unlike the motions of a butterfly. Doubtless the flight of all hirundines is influenced by, and adapted to, the peculiar sort of insects which furnish their food. Hence it would be worth inquiry to examine what particular genus of insects affords the principal food of each respective species of swallow.

Notwithstanding what has been advanced above, some few sand-martins, I see, haunt the skirts of London, frequenting the dirty pools in Saint George's-Fields, and about White-Chapel. The question is where these build, since there are no banks or bold shores in that neighbourhood: perhaps they nestle in the scaffold holes of some old or new deserted building. They dip and wash as they fly sometimes, like the house-martin and swallow.

Sand-martins differ from their congeners in the diminutiveness of their size, and in their colour, which is what is usually called a mouse-colour. Near Valencia, in Spain, they are taken, says Willughby, and sold in the markets for the table; and are called by the country people, probably from their desultory jerking manner of flight, *Papilion de Montagna.**

LETTER 21

Selborne, Sept. 28, 1774

Dear Sir,

As the swift or black-martin is the largest of the British hirundines, so is it undoubtedly the latest comer. For I remember but one instance of its appearing before the last week in April: and in some of our late frosty, harsh springs, it has not been seen till the beginning of May. This species usually arrives in pairs.

The swift, like the sand-martin, is very defective in architecture, making no crust, or shell, for its nest; but forming it of dry grasses and feathers, very rudely and inartificially put together. With all my attention to these birds, I have never been able once to discover one in the act of collecting or carrying in materials: so that I have suspected (since their nests are exactly the same) that they sometimes usurp upon the house-sparrows, and expel them, as sparrows do the house and sand-martin; well remembering that I have seen them squabbling together at the entrance of their holes; and the sparrows up in arms, and much-disconcerted at these intruders. And yet I am assured, by a nice observer in such matters, that they do collect feathers for their nests in Andalusia; and that he has shot them with such materials in their mouths.

Swifts, like sand-martins, carry on the business of nidification quite in the dark, in crannies of castles, and towers, and steeples, and upon the tops of the walls of churches under the roof; and therefore cannot be so narrowly watched as those species that build more openly: but, from what I could ever observe, they begin nesting about the middle of May; and I have remarked, from eggs taken, that they have sat hard by the ninth of June. In general they haunt tall build-ings, churches, and steeples, and breed only in such: yet in this village some pairs frequent the lowest and meanest cottages, and edu-cate their young under those thatched roofs. We remember but one instance where they breed out of buildings; and that is in the sides of a deep chalkpit near the town of Odiham, in this county, where we have seen many pairs entering the crevices, and skimming and squeaking round the precipices.

As I have regarded these amusive birds with no small attention, if I should advance something new and peculiar with respect to them, and different from all other birds, I might perhaps be credited; espe-cially as my assertion is the result of many years exact observation. The fact that I would advance is, that swifts tread, or copulate, on the wing: and I would wish any nice observer, that is startled at this sup-position, to use his own eyes, and I think he will soon be convinced. In another class of animals, *viz.* the insect, nothing is so common as to see the different species of many genera in conjunction as they fly. The swift is almost continually on the wing; and as it never settles on the ground, on trees, or roofs, would seldom find opportunity for amorous rites, was it not enabled to indulge them in the air. If any person would watch these birds of a fine morning in May, as they are sailing round at a great height from the ground, he would see, every now and then, one drop on the back of another, and both of them sink down together for many fathoms with a loud piercing shriek. This I take to be the juncture when the business of generation is carrying on.

As the swift eats, drinks, collects materials for its nest, and, as it seems, propagates on the wing; it appears to live more in the air than any other bird, and to perform all functions there save those of sleep-ing and incubation.

This *hirundo* differs widely from its congeners in laying invariably but two eggs at a time, which are milk-white, long, and peaked at the small end; whereas the other species lay at each brood from four to six. It is a most alert bird, rising very early, and retiring to roost very late; and is on the wing in the height of summer at least sixteen hours. In the longest days it does not withdraw to rest till a quarter before nine in the evening, being the latest of all day birds. Just before they retire whole groups of them assemble high in the air, and squeak, and shoot about with wonderful rapidity. But this bird is never so much alive as in sultry thundry weather, when it expresses great alacrity, and calls forth all its powers. In hot mornings several, getting together in little parties, dash round the steeples and churches, squeaking as they go in a very clamorous manner: these, by nice observers, are supposed to be males serenading their sitting hens; and not without reason, since they seldom squeak till they come close to the walls or eaves, and since those within utter at the same time a little inward note of complacency.

When the hen has sat hard all day, she rushes forth just as it is almost dark, and stretches and relieves her weary limbs, and snatches a scanty meal for a few minutes, and then returns to her duty of incubation. Swifts, when wantonly and cruelly shot while they have young, discover a little lump of insects in their mouths, which they pouch and hold under their tongue. In general they feed in a much higher district than the other species; a proof that gnats and other insects do also abound to a considerable height in the air: they also range to vast distances; since loco-motion is no labour to them, who are endowed with such wonderful powers of wing. Their powers seem to be in proportion to their levers; and their wings are longer in proportion than those of almost any other bird. When they mute,* or ease themselves in flight, they raise their wings, and make them meet over their backs.

At some certain times in the summer I had remarked that swifts were hawking very low for hours together over pools and streams; and could not help inquiring into the object of their pursuit that induced them to descend so much below their usual range. After some trouble, I found that they were taking *phryganeae*, *ephemerae*, and *libellulae* (cadew-flies, may-flies, and dragon-flies) that were just emerged out of their aurelia state. I then no longer wondered that they should be so willing to stoop for a prey that afforded them such plentiful and succulent nourishment.

They bring out their young about the middle or latter end of July: but as these never become perchers, nor, that ever I could discern, are fed on the wing by their dams, the coming forth of the young is not so notorious* as in the other species.

On the thirtieth of last June I untiled the eaves of an house where many pairs build,* and found in each nest only two squab, naked pulli: on the eighth of July I repeated the same inquiry, and found they had made very little progress towards a fledged state, but were still naked and helpless. From whence we may conclude that birds whose way of life keeps them perpetually on the wing would not be able to quit their nest till the end of the month. Swallows and martins, that have numerous families, are continually feeding them every two or three minutes; while swifts, that have but two young to maintain, are much at their leisure, and do not attend on their nests for hours together.

Sometimes they pursue and strike at hawks that come in their way; but not with that vehemence and fury that swallows express on the same occasion. They are out all day long in wet days, feeding about, and disregarding still rain:* from whence two things may be gathered; first, that many insects abide high in the air, even in rain; and next, that the feathers of these birds must be well preened to resist so much wet. Windy, and particularly windy weather with heavy showers, they dislike; and on such days withdraw, and are scarce ever seen.

There is a circumstance respecting the colour of swifts, which seems not to be unworthy our attention. When they arrive in the spring they are all over of a glossy, dark soot-colour, except their chins, which are white; but, by being all day long in the sun and air, they become quite weather-beaten and bleached before they depart, and yet they return glossy again in the spring. Now, if they pursue the sun into lower latitudes, as some suppose, in order to enjoy a perpetual summer, why do they not return bleached? Do they not rather perhaps retire to rest for a season, and at that juncture moult and change their feathers, since all other birds are known to moult soon after the season of breeding?*

Swifts are very anomalous in many particulars, dissenting from all their congeners not only in the number of their young, but in breeding but once in a summer; whereas all the other British hirundines breed invariably twice. It is past all doubt that swifts can breed but once, since they withdraw in a short time after the flight of their young, and some time before their congeners bring out their second broods. We may here remark, that, as swifts breed but once in a summer, and only two at a time, and the other hirundines twice, the latter, who lay from four to six eggs, increase at an average five times as fast as the former.

But in nothing are swifts more singular than in their early retreat. They retire, as to the main body of them, by the tenth of August, and sometimes a few days sooner: and every straggler invariably withdraws by the twentieth, while their congeners, all of them, stay till the beginning of October; many of them all through that month, and some occasionally to the beginning of November. This early retreat is mysterious and wonderful, since that time is often the sweetest season in the year. But, what is more extraordinary, they begin to retire still earlier in the most southerly parts of Andalusia, where they can be no ways influenced by any defect of heat; or, as one might suppose, defect of food. Are they regulated in their motions with us by a failure of food, or by a propensity to moulting, or by a disposition to rest after so rapid a life, or by what? This is one of those incidents in natural history that not only baffles our searches, but almost eludes our guesses!

These hirundines never perch on trees or roofs, and so never congregate with their congeners. They are fearless while haunting their nesting places, and are not to be scared with a gun; and are often beaten down with poles and cudgels as they stoop to go under the eaves. Swifts are much infested with those pests to the genus called *hippoboscae hirundinis*; and often wriggle and scratch themselves, in their flight, to get rid of that clinging annoyance.

Swifts are no songsters, and have only one harsh screaming note; yet there are ears to which it is not displeasing, from an agreeable association of ideas, since that note never occurs but in the most lovely summer weather.

They never settle on the ground but through accident; and when down can hardly rise, on account of the shortness of their legs and the length of their wings: neither can they walk, but only crawl; but they have a strong grasp with their feet, by which they cling to walls. Their bodies being flat they can enter a very narrow crevice; and where they cannot pass on their bellies they will turn up edgewise.

The particular formation of the foot discriminates the swift from all the British hirundines; and indeed from all other known birds, the *hirundo melba*, or great white-bellied swift of Gibraltar, excepted; for it is so disposed as to carry 'omnes quatuor digitos anticos' all its four toes forward; besides the least toe, which should be the back-toe, consists of one bone alone, and the other three only of two apiece. A construction most rare and peculiar, but nicely adapted to the purposes in which their feet are employed. This, and some peculiarities attending the nostrils and under mandible, have induced a discerning[1] naturalist to suppose that this species might constitute a genus *per se* [in itself].

In London a party of swifts frequents the Tower, playing and feeding over the river just below the bridge: others haunt some of the churches of the Borough next the fields; but do not venture, like the house-martin, into the close crowded part of the town.

The Swedes have bestowed a very pertinent name on this swallow, calling it *ring swala*, from the perpetual rings or circles that it takes round the scene of its nidification.

Swifts feed on coleoptera, or small beetles with hard cases over their wings, as well as on the softer insects; but it does not appear how they can procure gravel to grind their food, as swallows do, since they never settle on the ground. Young ones, over-run with *hippoboscae*, are sometimes found, under their nests, fallen to the ground; the number of vermin rendering their abode insupportable any longer. They frequent in this village several abject cottages; yet a succession still haunts the same unlikely roofs: a good proof this that the same birds return to the same spots. As they must stoop very low to get up under these humble eaves, cats lie in wait, and sometimes catch them on the wing.

On the fifth of July, 1775, I again untiled part of a roof over the nest of a swift. The dam sat in the nest; but so strongly was she affected by natural στοργή [affection] for her brood, which she supposed to be in danger, that, regardless of her own safety, she would not stir, but lay sullenly by them, permitting herself to be taken in hand. The squab young we brought down and placed on the grass-plot, where they tumbled about, and were as helpless as a new-born child. While we contemplated their naked bodies, their unwieldy disproportioned abdomina, and their heads, too heavy for their necks to support, we could not but wonder when we reflected that these shiftless beings in a little more than a fortnight would be able to dash through the air almost with the inconceivable swiftness of a meteor; and perhaps, in their emigration, must traverse vast continents and oceans as distant as the equator. So soon does Nature advance small birds to their ἡλικία [prime of life], or state of perfection; while the progressive growth of men and large quadrupeds is slow and tedious!*

I am, &c.

---

[1] John Antony Scopoli, of Carniola, M. D.

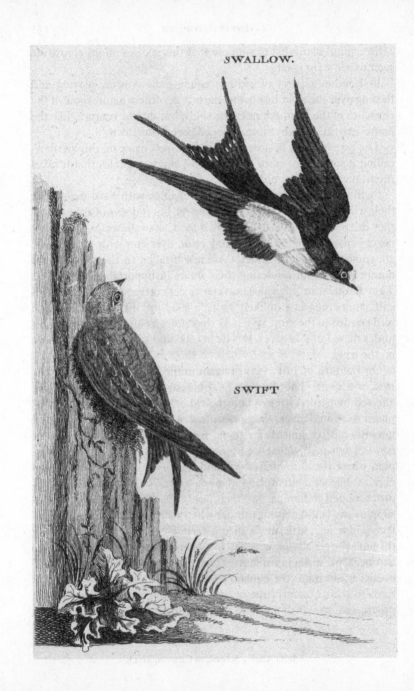

SWALLOW.

SWIFT

LETTER 22

Selborne, Sept. 13, 1774

Dear Sir,

By means of a straight cottage-chimney I had an opportunity this summer of remarking, at my leisure, how swallows ascend and descend through the shaft: but my pleasure, in contemplating the address with which this feat was performed to a considerable depth in the chimney, was somewhat interrupted by apprehensions lest my eyes might undergo the same fate with those of Tobit.[1]*

Perhaps it may be some amusement to you to hear at what times the different species of hirundines arrived this spring in three very distant counties of this kingdom. With us the swallow was seen first on April the 4th, the swift on April the 24th, the bank-martin on April the 12th, and the house-martin not till April the 30th. At South Zele, Devonshire, swallows did not arrive till April the 25th; swifts, in plenty, on May the 1st; and house-martins not till the middle of May. At Blackburn, in Lancashire, swifts were seen April the 28th, swallows April the 29th, house-martins May the 1st.* Do these different dates, in such distant districts, prove any thing for or against migration?

A farmer, near Weyhill, fallows his land with two teams of asses; one of which works till noon, and the other in the afternoon. When these animals have done their work, they are penned all night, like sheep, on the fallow. In the winter they are confined and foddered in a yard, and make plenty of dung.

Linnaeus says that hawks 'paciscuntur inducias cum avibus, quamdiu cuculus cuculat [agree an armistice with other birds for the period that the cuckoo calls]': but it appears to me that, during that period, many little birds are taken and destroyed by birds of prey, as may be seen by their feathers left in lanes and under hedges.

The missel-thrush is, while breeding, fierce and pugnacious, driving such birds as approach its nest, with great fury, to a distance. The Welch call it *pen y llwyn*,* the head or master of the coppice. He suffers no magpie, jay, or blackbird, to enter the garden where he haunts; and is, for the time, a good guard to the new-sown legumens.

---

[1] Tobit 2. 10.

In general he is very successful in the defence of his family: but once I observed in my garden, that several magpies came determined to storm the nest of a missel-thrush: the dams defended their mansion with great vigour, and fought resolutely *pro aris et facis* [for hearth and home. Cicero]; but numbers at last prevailed, they tore the nest to pieces, and swallowed the young alive.

In the season of nidification the wildest birds are comparatively tame. Thus the ring-dove breeds in my fields, though they are continually frequented; and the missel-thrush, though most shy and wild in the autumn and winter, builds in my garden close to a walk where people are passing all day long.

Wall-fruit abounds with me this year; but my grapes, that used to be forward and good, are at present backward beyond all precedent: and this is not the worst of the story; for the same ungenial weather, the same black cold solstice, has injured the more necessary fruits of the earth, and discoloured and blighted our wheat. The crop of hops promises to be very large.

Frequent returns of deafness incommode me sadly, and half disqualify me for a naturalist; for, when those fits are upon me, I lose all the pleasing notices and little intimations arising from rural sounds; and May is to me as silent and mute with respect to the notes of birds, &c. as August. My eyesight is, thank God, quick and good; but with respect to the other sense, I am, at times, disabled:

And wisdom at one entrance quite shut out.

[Milton, *Paradise Lost*, III. 50]

LETTER 23

Selborne, June 8, 1775

Dear Sir,

On September the 21st, 1741, being then on a visit, and intent on field-diversions, I rose before daybreak: when I came into the enclosures, I found the stubbles and clover-grounds matted all over with a thick coat of cobweb, in the meshes of which a copious and heavy dew hung so plentifully that the whole face of the country seemed, as it were, covered with two or three setting-nets drawn one over another.

When the dogs attempted to hunt, their eyes were so blinded and hoodwinked* that they could not proceed, but were obliged to lie down and scrape the incumbrances from their faces with their fore-feet, so that, finding my sport interrupted, I returned home musing in my mind on the oddness of the occurrence.

As the morning advanced the sun became bright and warm, and the day turned out one of those most lovely ones which no season but the autumn produces; cloudless, calm, serene, and worthy of the south of France itself.

About nine an appearance very unusual began to demand our attention, a shower of cobwebs falling from very elevated regions, and continuing, without any interruption, till the close of the day. These webs were not single filmy threads, floating in the air in all directions, but perfect flakes or rags; some near an inch broad, and five or six long, which fell with a degree of velocity that shewed they were considerably heavier than the atmosphere.

On every side as the observer turned his eyes might he behold a continual succession of fresh flakes falling into his sight, and twink-ling like stars as they turned their sides towards the sun.

How far this wonderful shower extended would be difficult to say; but we know that it reached Bradley, Selborne, and Alresford, three places which lie in a sort of a triangle, the shortest of whose sides is about eight miles in extent.

At the second of those places there was a gentleman (for whose veracity and intelligent turn we have the greatest veneration)* who observed it the moment he got abroad; but concluded that, as soon as he came upon the hill above his house, where he took his morning rides, he should be higher than this meteor,* which he imagined might have been blown, like thistle-down, from the common above: but, to his great astonishment, when he rode to the most elevated part of the down, 300 feet above his fields, he found the webs in appear-ance still as much above him as before; still descending into sight in a constant succession, and twinkling in the sun, so as to draw the atten-tion of the most incurious.

Neither before nor after was any such fall observed; but on this day the flakes hung in the trees and hedges so thick, that a diligent person sent out might have gathered baskets full.

The remark that I shall make on these cobweb-like appearances, called gossamer, is, that, strange and superstitious as the notions about them were formerly,* nobody in these days doubts but that they are the real production of small spiders, which swarm in the fields in fine weather in autumn, and have a power of shooting out webs from their tails so as to render themselves buoyant, and lighter than air. But why these apterous* insects should *that day* take such a wonderful aerial excursion, and why their webs should at once become so gross and material as to be considerably more weighty than air, and to descend with precipitation, is a matter beyond my skill. If I might be allowed to hazard a supposition, I should imagine that those filmy threads, when first shot, might be entangled in the rising dew, and so drawn up, spiders and all, by a brisk evaporation into the regions where clouds are formed: and if the spiders have a power of coiling and thickening their webs in the air, as Dr Lister says they have, (see his *Letters* to Mr Ray) then, when they were become heavier than the air, they must fall.*

Every day in fine weather, in autumn chiefly, do I see those spiders shooting out their webs and mounting aloft: they will go off from your finger if you will take them into your hand. Last summer one alighted on my book as I was reading in the parlour; and, running to the top of the page, and shooting out a web, took its departure from thence. But what I most wondered at was, that it went off with considerable velocity in a place where no air was stirring; and I am sure that I did not assist it with my breath. So that these little crawlers seem to have, while mounting, some loco-motive power without the use of wings, and to move in the air faster than the air itself.

LETTER 24

Selborne, Aug. 15, 1775

Dear Sir,

There is a wonderful spirit of sociality in the brute creation, independent of sexual attachment: the congregating of gregarious birds in the winter is a remarkable instance.

Many horses, though quiet with company, will not stay one minute in a field by themselves: the strongest fences cannot restrain them. My neighbour's horse will not only not stay by himself abroad, but he will not bear to be left alone in a strange stable without discovering the utmost impatience, and endeavouring to break the rack and manger with his fore feet. He has been known to leap out at a stable-window, through which dung was thrown, after company; and yet in other respects is remarkably quiet. Oxen and cows will not fatten by themselves; but will neglect the finest pasture that is not recommended by society. It would be needless to instance in sheep, which constantly flock together.

But this propensity seems not to be confined to animals of the same species; for we know a doe, still alive, that was brought up from a little fawn with a dairy of cows; with them it goes a-field, and with them it returns to the yard. The dogs of the house take no notice of this deer, being used to her; but, if strange dogs come by, a chase ensues; while the master smiles to see his favourite securely leading her pursuers over hedge, or gate, or stile, till she returns to the cows, who, with fierce lowings and menacing horns, drive the assailants quite out of the pasture.

Even great disparity of kind and size does not always prevent social advances and mutual fellowship. For a very intelligent and observant person has assured me that, in the former part of his life, keeping but one horse, he happened also on a time to have but one solitary hen. These two incongruous animals spent much of their time together in a lonely orchard, where they saw no creature but each other. By degrees an apparent regard began to take place between these two sequestered individuals. The fowl would approach the quadruped with notes of complacency, rubbing herself gently against his legs: while the horse would look down with satisfaction, and move with the greatest caution and circumspection, lest he should trample on his dimunitive companion. Thus, by mutual good offices, each seemed to console the vacant hours of the other: so that Milton, when he puts the following sentiment in the mouth of Adam, seems to be somewhat mistaken:

> Much less can bird with beast, or fish with fowl,
> So well converse, nor with the ox the ape.

[Milton, *Paradise Lost*, VIII. 395–6]

I am, &c.

LETTER 25

Selborne, Oct. 2, 1775

Dear Sir,

We have two gangs or hordes of gypsies which infest the south and west of England, and come round in their circuit two or three times in the year. One of these tribes calls itself by the noble name of Stanley, of which I have nothing particular to say; but the other is distinguished by an appellative somewhat remarkable—As far as their harsh gibberish can be understood, they seem to say that the name of their clan is Curleople: now the termination of this word is apparently Grecian: and as Mezeray and the gravest historians all agree that these vagrants did certainly migrate from Egypt and the East, two or three centuries ago, and so spread by degrees over Europe, may not this family-name, a little corrupted, be the very name they brought with them from the Levant? It would be matter of some curiosity, could one meet with an intelligent person among them, to inquire whether, in their jargon, they still retain any Greek words: the Greek radicals will appear in hand, foot, head, water, earth, &c. It is possible that amidst their cant and corrupted dialect many mutilated remains of their native language might still be discovered.

With regard to those peculiar people, the gypsies, one thing is very remarkable, and especially as they came from warmer climates; and that is, that while other beggars lodge in barns, stables, and cow-houses, these sturdy savages seem to pride themselves in braving the severities of winter, and in living *sub dio* [in the open air] the whole year round. Last September was as wet a month as ever was known; and yet during those deluges did a young gypsy-girl lie-in* in the midst of one of our hop-gardens, on the cold ground, with nothing over her but a piece of a blanket extended on a few hazel-rods bent hoop fashion, and stuck into the earth at each end, in circumstances too trying for a cow in the same condition: yet within this garden there was a large hop-kiln, into the chambers of which she might have retired, had she thought shelter an object worthy her attention.

Europe itself, it seems, cannot set bounds to the rovings of these vagabonds; for Mr Bell, in his return from Peking, met a gang of these people on the confines of Tartary, who were endeavouring to penetrate those deserts and try their fortune in China.[1]

Gypsies are called in French, *Bohemiens*; in Italian and modern Greek, *Zingani*.

I am, &c.

LETTER 26

Selborne, Nov. 1, 1775

Dear Sir,

> Hîc . . . taedae pingues, hîc plurimus ignis
> Semper, et affiduâ postes fuligine nigri.

> [Here, with the pine-wood torches full of burning fat and
> the fire always blazing high, the door-posts are black
> from the unremitting soot.]   [Virgil, *Eclogues* 7]

I shall make no apology for troubling you with the detail of a very simple piece of domestic oeconomy, being satisfied that you think nothing beneath your attention that tends to utility: the matter alluded to is the use of rushes instead of candles, which I am well aware prevails in many districts besides this; but as I know there are countries also where it does not obtain, and as I have considered the subject with some degree of exactness, I shall proceed in my humble story, and leave you to judge of the expediency.

The proper species of rush for this purpose seems to be the *juncus conglomeratus*, or common soft rush, which is to be found in most moist pastures, by the sides of streams, and under hedges. These rushes are in best condition in the height of summer; but may be gathered, so as to serve the purpose well, quite on to autumn. It would be needless to add that the largest and longest are best. Decayed* labourers, women, and children, make it their business to procure and prepare them. As soon as they are cut they must be flung into water, and kept there; for otherwise they will dry and shrink, and the peel will not run. At first a person would find it no easy matter to divest a rush of its peel or rind, so as to leave one regular, narrow,

---

[1] See Bell's *Travels in China.**

even rib from top to bottom that may support the pith: but this, like other feats, soon becomes familiar even to children; and we have seen an old woman, stone-blind, performing this business with great dispatch, and seldom failing to strip them with the nicest regularity. When these *junci* are thus far prepared, they must lie out on the grass to be bleached, and take the dew for some nights, and afterwards be dried in the sun.

Some address is required in dipping these rushes in the scalding fat or grease; but this knack also is to be attained by practice. The careful wife of an industrious Hampshire labourer obtains all her fat for nothing; for she saves the scummings of her bacon-pot for this use; and, if the grease abounds with salt, she causes the salt to precipitate to the bottom, by setting the scummings in a warm oven. Where hogs are not much in use, and especially by the sea-side, the coarser animal-oils will come very cheap. A pound of common grease may be procured for four pence; and about fix pounds of grease will dip a pound of rushes; and one pound of rushes may be bought for one shilling: so that a pound of rushes, medicated* and ready for use, will cost three shillings. If men that keep bees will mix a little wax with the grease, it will give it a consistency, and render it more cleanly, and make the rushes burn longer: mutton-suet would have the same effect.

A good rush, which measured in length two feet four inches and an half, being minuted,* burnt only three minutes short of an hour: and a rush still of greater length has been known to burn one hour and a quarter.

These rushes give a good clear light. Watch-lights (coated with tallow), it is true, shed a dismal one, 'darkness visible;'* but then the wick of those have two ribs of the rind, or peel, to support the pith, while the wick of the dipped rush has but one. The two ribs are intended to impede the progress of the flame and make the candle last.

In a pound of dry rushes, avoirdupois, which I caused to be weighed and numbered, we found upwards of one thousand six hundred individuals. Now suppose each of these burns, one with another, only half an hour, then a poor man will purchase eight hundred hours of light, a time exceeding thirty-three entire days, for three shillings. According to this account each rush, before dipping, costs ⅓₃ of a farthing,* and ¹⁄₁₁ afterwards. Thus a poor family will enjoy 5½ hours of comfortable light for a farthing. An experienced old housekeeper assures me that one pound and an half of rushes completely supplies his family the year round, since working people burn no candle in the long days, because they rise and go to bed by daylight.

Little farmers use rushes much in the short days, both morning and evening, in the dairy and kitchen; but the very poor, who are always the worst oeconomists, and therefore must continue very poor, buy an halfpenny candle every evening, which, in their blowing open rooms, does not burn much more than two hours. Thus have they only two hours light for their money instead of eleven.

While on the subject of rural oeconomy, it may not be improper to mention a pretty implement of housewifery that we have seen no where else; that is, little neat besoms which our foresters make from the stalks of the *polytricum commune*, or great golden maiden-hair, which they call silk-wood, and find plenty in the bogs. When this moss is well combed and dressed, and divested of its outer skin, it becomes of a beautiful bright-chesnut colour; and, being soft and pliant, is very proper for the dusting of beds, curtains, carpets, hangings, &c. If these besoms were known to the brushmakers in town, it is probable they might come much in use for the purpose above-mentioned.[1]

I am, &c.

LETTER 27

Selborne, Dec. 12, 1775

Dear Sir,

We had in this village more than twenty years ago an idiot-boy, whom I well remember, who, from a child, shewed a strong propensity to bees; they were his food, his amusement, his sole object. And as people of this cast have seldom more than one point in view, so this lad exerted all his few faculties on this one pursuit. In the winter he dozed away his time, within his father's house, by the fire side, in a kind of torpid state, seldom departing from the chimney-corner; but in the summer he was all alert, and in quest of his game in the fields,

---

[1] A besom of this sort is to be seen in Sir Ashton Lever's Museum.

and on sunny banks. Honey-bees, humble-bees, and wasps, were his prey wherever he found them: he had no apprehensions from their stings, but would seize them *nudis manibus* [with bare hands], and at once disarm them of their weapons, and suck their bodies for the sake of their honey-bags. Sometimes he would fill his bosom between his shirt and his skin with a number of these captives; and sometimes would confine them in bottles. He was a very *merops apiaster*, or bee-bird; and very injurious to men that kept bees; for he would slide into their bee-gardens, and, sitting down before the stools,* would rap with his finger on the hives, and so take the bees as they came out. He has been known to overturn hives for the sake of honey, of which he was passionately fond. Where metheglin* was making he would linger round the tubs and vessels, begging a draught of what he called bee-wine. As he ran about he used to make a humming noise with his lips, resembling the buzzing of bees. This lad was lean and sallow, and of a cadaverous complexion; and, except in his favourite pursuit, in which he was wonderfully adroit, discovered no manner of under-standing. Had his capacity been better, and directed to the same object, he had perhaps abated much of our wonder at the feats of a more modern exhibiter of bees: and we may justly say of him now,

> Thou,
> Had thy presiding star propitious shone,
> Should'st *Wildman* be . . .*

When a tall youth he was removed from hence to a distant village, where he died, as I understand, before he arrived at manhood.

I am, &c.

LETTER 28

Selborne, Jan. 8, 1776

Dear Sir,

It is the hardest thing in the world to shake off superstitious prejudices: they are sucked in as it were with our mother's milk; and, growing up with us at a time when they take the fastest hold and make the most lasting impressions, become so interwoven into our very constitutions, that the strongest good sense is required to disengage ourselves from them. No wonder therefore that the lower people retain them their whole lives through, since their minds are not invigorated by a liberal education, and therefore not enabled to make any efforts adequate to the occasion.

Such a preamble seems to be necessary before we enter on the superstitions of this district, lest we should be suspected of exaggeration in a recital of practices too gross for this enlightened age.

But the people of Tring, in Hertfordshire, would do well to remember, that no longer ago than the year 1751, and within twenty miles of the capital, they seized on two superannuated wretches, crazed with age, and overwhelmed with infirmities, on a suspicion of witchcraft; and, by trying experiments, drowned them in a horse-pond.

In a farm-yard near the middle of this village stands, at this day, a row of pollard-ashes, which, by the seams and long cicatrices* down their sides, manifestly shew that, in former times, they have been cleft asunder. These trees, when young and flexible, were severed and held open by wedges, while ruptured children, stripped naked, were pushed through the apertures, under a persuasion that, by such a process, the poor babes would be cured of their infirmity. As soon as the operation was over, the tree, in the suffering part, was plastered with loam, and carefully swathed up. If the parts coalesced and soldered together, as usually fell out, where the feat was performed with any adroitness at all, the party was cured; but, where the cleft continued to gape, the operation, it was supposed, would prove ineffectual. Having occasion to enlarge my garden not long since, I cut down two or three such trees, one of which did not grow together.

We have several persons now living in the village, who, in their childhood, were supposed to be healed by this superstitious ceremony, derived down perhaps from our Saxon ancestors, who practised it before their conversion to Christianity.

At the south corner of the Plestor, or area, near the church, there stood, about twenty-years ago, a very old grotesque hollow pollard-ash, which for ages had been looked on with no small veneration as a shrew-ash. Now a shrew-ash is an ash whose twigs or branches, when gently applied to the limbs of cattle, will immediately relieve the pains which a beast suffers from the running of a shrew-mouse over the part affected: for it is supposed that a shrew-mouse is of so baneful and deleterious a nature, that wherever it creeps over a beast, be it horse, cow, or sheep, the suffering animal is afflicted with cruel anguish, and threatened with the loss of the use of the limb. Against this accident, to which they were continually liable, our provident fore-fathers always kept a shrew-ash at hand, which, when once medicated, would maintain its virtue for ever. A shrew-ash was made thus[1]:—Into the body of the tree a deep hole was bored with an auger, and a poor devoted shrew-mouse was thrust in alive, and plugged in, no doubt, with several quaint incantations long since forgotten. As the ceremonies necessary for such a consecration are no longer understood, all succession is at an end, and no such tree is known to subsist in the manor, or hundred.*

As to that on the Plestor

The late vicar stubb'd and burnt it.*

when he was way-warden,* regardless of the remonstrances of the by-standers, who interceded in vain for its preservation, urging its power and efficacy, and alledging that it had been

Religione patrum multos servata per annos
[Saved for many years by the reverence of our
ancestors.]   [Virgil, *Aeneid* 2]

I am, &c.

LETTER 29

Selborne, Feb. 7, 1776

Dear Sir,

In heavy fogs, on elevated situations especially, trees are perfect alembics;* and no one that has not attended to such matters can imagine how much water one tree will distil in a night's time, by condensing the vapour, which trickles down the twigs and boughs, so as to make the ground below quite in a float. In Newton-lane, in October 1775, on a misty day, a particular oak in leaf dropped so fast that the cart-way stood in puddles and the ruts ran with water, though the ground in general was dusty.

In some of our smaller islands in the West-Indies, if I mistake not,

[1] For a similar practice, see Plot's *Staffordshire*.

there are no springs or rivers; but the people are supplied with that necessary element, water, merely by the dripping of some large tall trees, which, standing in the bosom of a mountain, keep their heads constantly enveloped with fogs and clouds, from which they dispense their kindly never-ceasing moisture; and so render those districts habitable by condensation alone.

Trees in leaf have such a vast proportion more of surface than those that are naked, that, in theory, their condensations should greatly exceed those that are stripped of their leaves; but, as the former imbibe also a great quantity of moisture, it is difficult to say which drip most: but this I know, that deciduous trees that are entwined with much ivy seem to distil the greatest quantity. Ivy-leaves are smooth, and thick, and cold, and therefore condense very fast; and besides ever-greens imbibe very little. These facts may furnish the intelligent with hints concerning what sorts of trees they should plant round small ponds that they would wish to be perennial; and shew them how advantageous some trees are in preference to others.

Trees perspire profusely, condense largely, and check evaporation so much, that woods are always moist; no wonder therefore that they contribute much to pools and streams.

That trees are great promoters of lakes and rivers appears from a well known fact in North-America; for, since the woods and forests have been grubbed and cleared, all bodies of water are much diminished; so that some streams, that were very considerable a century ago, will not now drive a common mill.[1] Besides, most woodlands, forests, and chases, with us abound with pools and morasses; no doubt for the reason given above.

To a thinking mind few phenomena are more strange than the state of little ponds on the summits of chalk-hills, many of which are never dry in the most trying droughts of summer. On chalk-hills I say, because in many rocky and gravelly soils springs usually break out pretty high on the sides of elevated grounds and mountains; but no person acquainted with chalky districts will allow that they ever saw springs in such a soil but in vallies and bottoms, since the waters of so pervious a stratum as chalk all lie on one dead level, as well-diggers have assured me again and again.

---

[1] Vide Kalm's *Travels to North-America* [trans. J. R. Foster, 2d ed., 1772].

Now we have many such little round ponds in this district; and one in particular on our sheep-down, three hundred feet above my house; which, though never above three feet deep in the middle, and not more than thirty feet in diameter, and containing perhaps not more than two or three hundred hogsheads of water, yet never is known to fail, though it affords drink for three hundred or four hundred sheep, and for at least twenty head of large cattle beside. This pond, it is true, is over-hung with two moderate beeches, that, doubtless, at times afford it much supply: but then we have others as small, that, without the aid of trees, and in spite of evaporation from sun and wind, and perpetual consumption by cattle, yet constantly maintain a moderate share of water, without overflowing in the wettest seasons, as they would do if supplied by springs. By my journal of May, 1775, it appears that 'the small and even considerable ponds in the vales are now dried up, while the small ponds on the very tops of hills are but little affected.' Can this difference be accounted for from evaporation alone, which certainly is more prevalent in bottoms? or rather have not these elevated pools some unnoticed recruits, which in the night time counterbalance the waste of the day; without which the cattle alone must soon exhaust them? And here it will be necessary to enter more minutely into the cause. Dr Hales, in his *Vegetable Statics*, advances, from experiment, that 'the moister the earth is the more dew falls on it in a night: and more than a double quantity of dew falls on a surface of water than there does on an equal surface of moist earth.'* Hence we see that water, by its coolness, is enabled to assimilate to itself a large quantity of moisture nightly by condensation; and that the air, when loaded with fogs and vapours, and even with copious dews, can alone advance a considerable and never-failing resource. Persons that are much abroad, and travel early and late; such as shepherds, fishermen, &c. can tell what prodigious fogs prevail in the night on elevated downs, even in the hottest parts of summer; and how much the surfaces of things are drenched by those swimming vapours, though, to the senses, all the while, little moisture seems to fall.

I am, &c.

LETTER 30

Selborne, April 3, 1776

Dear Sir,

Monsieur Herissant, a French anatomist, seems persuaded that he has discovered the reason why cuckoos do not hatch their own eggs; the impediment, he supposes, arises from the internal structure of their parts, which incapacitates them for incubation. According to this gentleman, the crop, or craw, of a cuckoo does not lie before the sternum at the bottom of the neck, as in the gallinae; columbae, &c. but immediately behind it, on and over the bowels, so as to make a large protuberance in the belly.[1]*

Induced by this assertion, we procured a cuckoo; and, cutting open the breast-bone, and exposing the intestines to sight, found the crop lying as mentioned above. This stomach was large and round, and stuffed hard like a pincushion with food, which, upon nice examination, we found to consist of various insects; such as small scarabs, spiders, and dragon-flies; the last of which we have seen cuckoos catching on the wing as they were just emerging out of the aurelia state. Among this farrago also were to be seen maggots, and many seeds, which belonged either to gooseberries, currants, cranberries, or some such fruit; so that these birds apparently subsist on insects and fruits: nor was there the least appearance of bones, feathers, or fur, to support the idle notion of their being birds of prey.

The sternum in this bird seemed to us to be remarkably short, between which and the anus lay the crop, or craw, and immediately behind that the bowels against the back-bone.

It must be allowed, as this anatomist observes, that the crop placed just upon the bowels must, especially when full, be in a very uneasy situation during the business of incubation; yet the test will be to examine whether birds that are actually known to sit for certain are not formed in a similar manner. This inquiry I proposed to myself to make with a fern-owl, or goat-sucker, as soon as opportunity offered: because, if their formation proves the same, the reason for incapacity in the cuckoo will be allowed to have been taken up somewhat hastily.

[1] *Histoire de l'Academie Royale,* 1752.

Not long after a fern-owl was procured, which, from its habit and shape, we suspected might resemble the cuckoo in its internal construction. Nor were our suspicions ill-grounded; for, upon the dissection, the crop, or craw, also lay behind the sternum, immediately on the viscera, between them and the skin of the belly. It was bulky, and stuffed hard with large *phalaenae*, moths of several sorts, and their eggs, which no doubt had been forced out of those insects by the action of swallowing.

Now as it appears that this bird, which is so well known to practise incubation, is formed in a similar manner with cuckoos, Monsieur Herissant's conjecture, that cuckoos are incapable of incubation from the disposition of their intestines, seems to fall to the ground; and we are still at a loss for the cause of that strange and singular peculiarity in the instance of the *cuculus canorus*.

We found the case to be the same with the ring-tail hawk, in respect to formation; and, as far as I can recollect, with the swift; and probably it is so with many more sorts of birds that are not granivorous.

I am, &c.

LETTER 31

Selborne, April 29, 1776

Dear Sir,

On August the 4th, 1775, we surprised a large viper, which seemed very heavy and bloated, as it lay in the grass basking in the sun.* When we came to cut it up, we found that the abdomen was crowded with young, fifteen in number; the shortest of which measured full seven inches, and were about the size of full-grown earth-worms. This little fry issued into the world with the true viper-spirit about them, shewing great alertness as soon as disengaged from the belly of the dam: they twisted and wriggled about, and set themselves up, and gaped very wide when touched with a stick, shewing manifest tokens of menace and defiance, though as yet they had no manner of fangs that we could find, even with the help of our glasses.

To a thinking mind nothing is more wonderful than that early instinct which impresses young animals with the notion of the situation of their natural weapons, and of using them properly in their own defence, even before those weapons subsist or are formed. Thus a young cock will spar at his adversary before his spurs are grown; and a calf or a lamb will push with their heads before their horns are sprouted. In the same manner did these young adders attempt to bite before their fangs were in being. The dam however was furnished with very formidable ones, which we lifted up (for they fold down when not used) and cut them off with the point of our scissars.

There was little room to suppose that this brood had ever been in the open-air before; and that they were taken in for refuge, at the mouth of the dam, when she perceived that danger was approaching; because then probably we should have found them somewhere in the neck, and not in the abdomen.

LETTER 32

Castration has a strange effect: it emasculates both man, beast, and bird, and brings them to a near resemblance of the other sex. Thus eunuchs have smooth unmuscular arms, thighs, and legs; and broad hips, and beardless chins, and squeaking voices. Gelt-stags and bucks have hornless heads, like hinds and does. Thus wethers have small horns, like ewes; and oxen large bent horns, and hoarse voices when they low, like cows: for bulls have short straight horns; and though they mutter and grumble in a deep tremendous tone, yet they low in a shrill high key. Capons have small combs and gills, and look pallid about the head, like pullets; they also walk without any parade, and hover chickens like hens. Barrow-hogs have also small tusks like sows.

Thus far it is plain that the deprivation of masculine vigour puts a stop to the growth of those parts or appendages that are looked upon as its insignia. But the ingenious Mr Lisle, in his book on husbandry, carries it much farther; for he says that the loss of those insignia alone has sometimes a strange effect on the ability itself: he had a boar so fierce and venereous, that, to prevent mischief, orders were given for his tusks to be broken off. No sooner had the beast suffered this injury than his powers forsook him, and he neglected those females to whom before he was passionately attached, and from whom no fences could restrain him.*

LETTER 33

The natural term of an hog's life is little known, and the reason is
plain—because it is neither profitable nor convenient to keep that
turbulent animal to the full extent of its time: however, my neighbour,
a man of substance, who had no occasion to study every little advantage to a nicety, kept an half bred Bantam-sow, who was as thick as she
was long, and whose belly swept on the ground till she was advanced
to her seventeenth year; at which period she shewed some tokens of
age by the decay of her teeth and the decline of her fertility.

For about ten years this prolific mother produced two litters in the
year of about ten at a time, and once above twenty at a litter; but, as
there were near double the number of pigs to that of teats, many died.
From long experience in the world this female was grown very sagacious and artful:—when she found occasion to converse* with a boar
she used to open all the intervening gates, and march, by herself, up
to a distant farm where one was kept; and when her purpose was
served would return by the same means. At the age of about fifteen
her litters began to be reduced to four or five; and such a litter she
exhibited when in her fatting-pen. She proved, when fat, good bacon,
juicy, and tender; the rind, or sward, was remarkably thin. At a moderate computation she was allowed to have been the fruitful parent of
three hundred pigs: a prodigious instance of fecundity in so large a
quadruped! She was killed in spring 1775.

                                                        I am, &c.

LETTER 34

                                        Selborne, May, 9, 1776
Dear Sir,

                    . . . admorunt ubera tigres.

                    [tigresses suckled (you).]

                          [Virgil, *Aeneid* 4]

We have remarked in a former letter how much incongruous animals, in a lonely state, may be attached to each other from a spirit of

sociality; in this it may not be amiss to recount a different motive which has been known to create as strange a fondness.

My friend had a little helpless leveret* brought to him, which the servants fed with milk in a spoon, and about the same time his cat kittened and the young were dispatched and buried. The hare was soon lost, and supposed to be gone the way of most fondlings, to be killed by some dog or cat. However, in about a fortnight, as the master was sitting in his garden in the dusk of the evening, he observed his cat, with tail erect, trotting towards him, and calling with little short inward notes of complacency, such as they use towards their kittens, and something gamboling after, which proved to be the leveret that the cat had supported with her milk, and continued to support with great affection.

Thus was a graminivorous animal nurtured by a carnivorous and predaceous one!

Why so cruel and sanguinary a beast as a cat, of the ferocious genus of *Feles*, the *murium leo*, as Linnaeus calls it, should be affected with any tenderness towards an animal which is its natural prey, is not so easy to determine.

This strange affection probably was occasioned by that desiderium,* those tender maternal feelings, which the loss of her kittens had awakened in her breast; and by the complacency and ease she derived to herself from the procuring her teats to be drawn, which were too much distended with milk, till, from habit, she became as much delighted with this foundling as if it had been her real offspring.

This incident is no bad solution of that strange circumstance which grave historians as well as the poets assert, of exposed children being sometimes nurtured by female wild beasts that probably had lost their young. For it is not one whit more marvellous that Romulus and Remus, in their infant state, should be nursed by a she-wolf, than that a poor little sucking leveret should be fostered and cherished by a bloody grimalkin.

> . . . viridi foetam Mavortis in antro
> Procubuisse lupam: geminos huic ubera circum
> Ludere pendentes pueros, et lambere matrem
> Impavidos: illam tereti cervice reflexam
> Mulcere alternos, et corpora singere linguâ.

[. . . the newly delivered she-wolf lying outstretched in the green cave
of Mars, the twin boys playing as they hung around her breasts lick-
ing their dam without fear, while she, bending back her shapely neck,
caressed them by turns and moulded their bodies with her tongue.]

[Virgil, *Aeneid* 8]

LETTER 35

Selborne, May 20, 1777

Dear Sir,

Lands that are subject to frequent inundations are always poor;
and probably the reason may be because the worms are drowned.
The most insignificant insects and reptiles are of much more consequence, and have much more influence in the oeconomy of Nature,
than the incurious are aware of; and are mighty in their effect, from
their minuteness, which renders them less an object of attention; and
from their numbers and fecundity. Earth-worms, though in appearance a small and despicable link in the chain of Nature, yet, if lost,
would make a lamentable chasm.* For, to say nothing of half the
birds, and some quadrupeds which are almost entirely supported by
them, worms seem to be the great promoters of vegetation, which
would proceed but lamely without them, by boring, perforating, and
loosening the soil, and rendering it pervious to rains and the fibres of
plants, by drawing straws and stalks of leaves and twigs into it; and,
most of all, by throwing up such infinite numbers of lumps of earth
called worm-casts, which, being their excrement, is a fine manure for
grain and grass. Worms probably provide new soil for hills and slopes
where the rain washes the earth away; and they affect slopes, probably to avoid being flooded. Gardeners and farmers express their
detestation of worms; the former because they render their walks
unsightly, and make them much work: and the latter because, as they
think, worms eat their green corn. But these men would find that the
earth without worms would soon become cold, hard-bound, and void
of fermentation; and consequently steril: and besides, in favour of
worms, it should be hinted that green corn, plants, and flowers, are not
so much injured by them as by many species of coleoptera (scarabs),
and *tipulae* (long-legs) in their larva, or grub-state; and by unnoticed

myriads of small shell-less snails, called slugs, which silently and imperceptibly make amazing havoc in the field and garden.[1]

These hints we think proper to throw out in order to set the inquisitive and discerning to work.

A good monography of worms would afford much entertainment and information at the same time, and would open a large and new field in natural history. Worms work most in the spring; but by no means lie torpid in the dead months; are out every mild night in the winter, as any person may be convinced that will take the pains to examine his grass-plots with a candle; are hermaphrodites, and much addicted to venery, and consequently very prolific.

I am, &c.

LETTER 36

Selborne, Nov. 22, 1777

Dear Sir,

You cannot but remember that the twenty-sixth and twenty-seventh of last March were very hot days; so sultry that every body complained and were restless under those sensations to which they had not been reconciled by gradual approaches.

This sudden summer-like heat was attended by many summer coincidences; for on those two days the thermometer rose to sixty-six in the shade; many species of insects revived and came forth; some bees swarmed in this neighbourhood; the old tortoise, near Lewes, in Sussex, awakened and came forth out of its dormitory; and, what is most to my present purpose, many house-swallows appeared and were very alert in many places, and particularly at Cobham, in Surrey.

But as that short warm period was succeeded as well as preceded by harsh severe weather, with frequent frosts and ice, and cutting winds, the insects withdrew, the tortoise retired again into the ground, and the swallows were seen no more until the tenth of April, when, the rigour of the spring abating, a softer season began to prevail.

Again; it appears by my journals for many years past that

---

[1] Farmer Young, of Norton-farm, says that this spring (1777) about four acres of his wheat in one field was entirely destroyed by slugs, which swarmed on the blades of corn, and devoured it as fast as it sprang.

house-martins retire, to a bird, about the beginning of October; so that a person not very observant of such matters would conclude that they had taken their last farewell: but then it may be seen in my diaries also that considerable flocks have discovered themselves again in the first week of November, and often on the fourth day of that month only for one day; and that not as if they were in actual migration, but playing about at their leisure and feeding calmly, as if no enterprize of moment at all agitated their spirits. And this was the case in the beginning of this very month; for, on the fourth of November, more than twenty house-martins, which, in appearance, had all departed about the seventh of October, were seen again, for that *one morning only*, sporting between my fields and the Hanger, and feasting on insects which swarmed in that sheltered district. The preceding day was wet and blustering, but the fourth was dark and mild, and soft, the wind at south-west, and the thermometer at 58'½; a pitch not common at that season of the year. Moreover, it may not be amiss to add in this place, that whenever the thermometer is above 50 the bat comes flitting out in every autumnal and winter-month.

From all these circumstances laid together, it is obvious that torpid insects, reptiles, and quadrupeds, are awakened from their profoundest slumbers by a little untimely warmth; and therefore that nothing so much promotes this death-like stupor as a defect of heat. And farther, it is reasonable to suppose that two whole species, or at least many individuals of those two species, of British hirundines, do never leave this island at all, but partake of the same benumbed state: for we cannot suppose that, after a month's absence, house-martins can return from southern regions to appear for one morning in November, or that house-swallows should leave the districts of Africa to enjoy, in March, the transient summer of a couple of days.

I am, &c.

LETTER 37

Selborne, Jan. 8, 1778

Dear Sir,

There was in this village several years ago a miserable pauper, who, from his birth, was afflicted with a leprosy, as far as we are aware of a singular kind, since it affected only the palms of his hands and the soles of his feet. This scaly eruption usually broke out twice in the year, at the spring and fall; and, by peeling away, left the skin so thin and tender that neither his hands or feet were able to perform their functions; so that the poor object was half his time on crutches, incapable of employ, and languishing in a tiresome state of indolence and inactivity. His habit was lean, lank, and cadaverous. In this sad plight he dragged on a miserable existence, a burden to himself and his parish, which was obliged to support him till he was relieved by death at more than thirty years, of age.

The good women, who love to account for every defect in children by the doctrine of longing, said that his mother felt a violent propensity for oysters, which she was unable to gratify; and that the black rough scurf on his hands and feet were the shells of that fish.* We knew his parents, neither of which were lepers; his father in particular lived to be far advanced in years.

In all ages the leprosy has made dreadful havock among mankind. The Israelites seem to have been greatly afflicted with it from the most remote times; as appears from the peculiar and repeated injunctions given them in the Levitical law.[1] Nor was the rancour of this foul disorder much abated in the last period of their commonwealth, as may be seen in many passages of the New Testament.

Some centuries ago this horrible distemper prevailed all Europe over; and our forefathers were by no means exempt, as appears by the large provision made for objects labouring under this calamity. There was an hospital for female lepers in the diocese of Lincoln, a noble one near Durham, three in London and Southwark, and perhaps many more in or near our great towns and cities. Moreover, some crowned heads, and other wealthy and charitable personages, bequeathed large legacies to such poor people as languished under this hopeless infirmity.

It must therefore, in these days, be, to an humane and thinking person, a matter of equal wonder and satisfaction, when he contemplates how nearly this pest is eradicated, and observes that a leper now is a rare sight. He will, moreover, when engaged in such a train of thought, naturally inquire for the reason. This happy change perhaps may have originated and been continued from the much smaller

---

[1] See Leviticus, chap. xiii and xiv.

quantity of salted meat and fish now eaten in these kingdoms; from the use of linen next the skin; from the plenty of better bread; and from the profusion of fruits, roots, legumes, and greens, so common in every family. Three or four centuries ago, before there were any enclosures, sown-grasses, field-turnips, or field-carrots, or hay, all the cattle which had grown fat in summer, and were not killed for winter-use, were turned out soon after Michaelmas to shift as they could through the dead months; so that no fresh meat could be had in winter or spring. Hence the marvellous account of the vast stores of salted flesh found in the larder of the eldest Spencer[1] in the days of Edward the Second, even so late in the spring as the third of May.* It was from magazines like these that the turbulent barons supported in idleness their riotous swarms of retainers ready for any disorder or mischief. But agriculture is now arrived at such a pitch of perfection, that our best and fattest meats are killed in the winter; and no man need eat salted flesh, unless he prefers it, that has money to buy fresh.

One cause of this distemper might be, no doubt, the quantity of wretched fresh and salt fish consumed by the commonalty at all seasons as well as in lent; which our poor now would hardly be persuaded to touch.

The use of linen changes, shirts or shifts, in the room of sordid and filthy woollen, long worn next the skin, is a matter of neatness comparatively modern; but must prove a great means of preventing cutaneous ails. At this very time woollen instead of linen prevails among the poorer Welch, who are subject to foul eruptions.

The plenty of good wheaten bread that now is found among all ranks of people in the south, instead of that miserable sort which used in old days to be made of barley or beans, may contribute not a little to the sweetening their blood and correcting their juices;* for the inhabitants of mountainous districts, to this day, are still liable to the itch and other cutaneous disorders, from a wretchedness and poverty of diet.

As to the produce of a garden, every middle-aged person of observation may perceive, within his own memory, both town and country, how vastly the consumption of vegetables is increased. Green-stalls in cities now support multitudes in a comfortable state, while gardeners get fortunes.* Every decent labourer also has his

---

[1] *Viz.* Six hundred bacons, eighty carcasses of beef, and six hundred muttons.

garden, which is half his support, as well as his delight; and common farmers provide plenty of beans, peas, and greens, for their hinds* to eat with their bacon; and those few that do not are despised for their sordid parsimony, and looked upon as regardless of the welfare of their dependants. Potatoes have prevailed in this little district, by means of premiums, within these twenty years only; and are much esteemed here now by the poor, who would scarce have ventured to taste them in the last reign.*

Our Saxon ancestors certainly had some sort of cabbage, because they call the month of February sprout-cale; but, long after their days, the cultivation of gardens was little attended to. The religious, being men of leisure, and keeping up a constant correspondence with Italy, were the first people among us that had gardens and fruit-trees in any perfection, within the walls of their abbies[1] and priories. The barons neglected every pursuit that did not lead to war or tend to the pleasure of the chase.

It was not till gentlemen took up the study of horticulture them-selves that the knowledge of gardening made such hasty advances. Lord Cobham, Lord Ila and Mr Waller of Beaconsfield, were some of the first people of rank that promoted the elegant science of orna-menting without despising the superintendence of the kitchen quar-ters and fruit walls.

A remark made by the excellent Mr Ray in his Tour of Europe* at once surprises us, and corroborates what has been advanced above; for we find him observing, so late as his days, that 'the Italians use several herbs for sallets,* which are not yet or have not been but lately used in England, *viz. selleri* (celery) which is nothing else but the sweet smallage; the young shoots whereof, with a little of the head of the root cut off, they eat raw with oil and pepper.' And farther he adds 'curled endive blanched is much used beyond seas; and, for a raw sallet, seemed to excel lettuce itself.' Now this journey was undertaken no longer ago than in the year 1663.

I am, &c.

---

[1] 'In monasteries the lamp of knowledge continued to burn, however dimly. In them men of business were formed for the state: the art of writing was cultivated by the monks; they were the only proficients in mechanics, gardening, and architecture.' See Dalrymple's *Annals of Scotland* [1776, 1779].

LETTER 38

Fortè puer, comitum seductus ab agmine fido,
Dixerat, ecquis adest? et, adest, responderat echo.
Hic stupet; utque àciem partes divisit in omnes;
Voce, veni, clamat magnâ. Vocat illa vocantem.

[Perchance a boy, parted from his trusty band of comrades, had
exclaimed, 'Is anyone here?' to which Echo replied, 'Anyone here'.
In amazement, and turning his keen glance in all directions, the
boy calls out in a loud voice, 'Come!' And Echo calls as he calls.]

[Ovid, *Metamorphoses* 3]

Selborne, Feb. 12, 1778

Dear Sir,

In a district so diversified as this, so full of hollow vales and hanging woods, it is no wonder that echoes should abound. Many we have discovered that return the cry of a pack of dogs, the notes of a hunting-horn, a tunable ring of bells, or the melody of birds, very agreeably; but we were still at a loss for a polysyllabical, articulate echo, till a young gentleman, who had parted from his company in a summer evening walk, and was calling after them, stumbled upon a very curious one in a spot where it might least be expected. At first he was much surprised, and could not be persuaded but that he was mocked by some boy; but, repeating his trials in several languages, and finding his respondent to be a very adroit polyglot, he then discerned the deception.

This echo in an evening, before rural noises cease, would repeat ten syllables most articulately and distinctly, especially if quick dactyls* were chosen. The last syllables of

Tityre, tu patulae recubans . . .

[You, O Tityrus, reclining . . .]

[Virgil, *Eclogues* 1]

were as audibly and intelligibly returned as the first: and there is no doubt, could trial have been made, but that at midnight, when the air is very elastic, and a dead stillness prevails, one or two syllables more might have been obtained; but the distance rendered so late an experiment very inconvenient.

Quick dactyls, we observed, succeeded best; for when we came to try its powers in slow, heavy, embarassed spondees* of the same number of syllables,

> Monstrum horrendum, informe, ingens . . .
>
> [A monster, horrible, misshapen, huge . . .]
>
> [Virgil, *Aeneid* 3]

we could perceive a return but of four or five.

All echoes have some one place to which they are returned stronger and more distinct than to any other; and that is always the place that lies at right angles with the object of repercussion, and is not too near, nor too far off. Buildings, or naked rocks, re-echo much more articulately than hanging wood or vales; because in the latter the voice is as it were entangled, and embarassed in the covert, and weakened in the rebound.

The true object of this echo, as we found by various experiments, is the stone-built, tiled hop-kiln in Gally-lane, which measures in front 40 feet, and from the ground to the eaves 12 feet. The true *centrum phonicum*,* or just distance, is one particular spot in the Kings-field, in the path to Nore-hill, on the very brink of the steep balk above the hollow cart way. In this case there is no choice of distance; but the path, by meer contingency, happens to be the lucky, the identical spot, because the ground rises or falls so immediately, if the speaker either retires or advances, that his mouth would at once be above or below the object.

We measured this polysyllabical echo with great exactness, and found the distance to fall very short of Dr Plot's rule for distinct articulation: for the Doctor, in his history of Oxfordshire, allows 120 feet for the return of each syllable distinctly: hence this echo, which gives ten distinct syllables, ought to measure 400 yards, or 120 feet to each syllable; whereas our distance is only 258 yards, or near 75 feet, to each syllable. Thus our measure falls short of the Doctor's, as five to eight: but then it must be acknowledged that this candid philosopher was convinced afterwards, that some latitude must be admitted of in the distance of echoes according to time and place.

When experiments of this sort are making, it should always be remembered that weather and the time of day have a vast influence on an echo; for a dull, heavy, moist air deadens and clogs the sound; and hot sunshine renders the air thin and weak, and deprives it of all its springiness; and a ruffling wind quite defeats the whole. In a still, clear, dewy evening the air is most elastic; and perhaps the later the hour the more so.

Echo has always been so amusing to the imagination, that the poets have personified her; and in their hands she has been the occasion of many a beautiful fiction. Nor need the gravest man be ashamed to appear taken with such a phaenomenon, since it may become the subject of philosophical or mathematical inquiries.

One should have imagined that echoes, if not entertaining, must at least have been harmless and inoffensive; yet Virgil advances a strange notion, that they are injurious to bees. After enumerating some probable and reasonable annoyances, such as prudent owners would wish far removed from their bee-gardens, he adds

> . . . aut ubi concava pulsu
> Saxa sonant, vocisque offensa resultat imago.
>
> [or where the arching rocks vibrate when struck,
> and the echo of a voice, thrown back, resounds.]
>
> [Virgil, *Georgics* 4]

This wild and fanciful assertion will hardly be admitted by the philosophers of these days; especially as they all now seem agreed that insects are not furnished with any organs of hearing at all. But if it should be urged, that though they cannot hear yet perhaps they may feel the repercussions of sounds, I grant it is possible they may. Yet that these impressions are distasteful or hurtful, I deny, because bees, in good summers, thrive well in my outlet, where the echoes are very strong: for this village is another Anathoth,* a place of responses or echoes. Besides, it does not appear from experiment that bees are in any way capable of being affected by sounds: for I have often tried my own with a large speaking-trumpet held close to their hives, and with such an exertion of voice as would have haled a ship at the distance of a mile, and still these insects pursued their various employments undisturbed, and without shewing the least sensibility or resentment.

Some time since its discovery this echo is become totally silent, though the object, or hop-kiln, remains: nor is there any mystery in this defect; for the field between is planted as an hop-garden, and the voice of the speaker is totally absorbed and lost among the poles and entangled foliage of the hops. And when the poles are removed in autumn the disappointment is the same; because a tall quick-set* hedge, nurtured up for the purpose of shelter to the hop ground, entirely interrupts the impulse and repercussion of the voice: so that till those obstructions are removed no more of its garrulity can be expected.

Should any gentleman of fortune think an echo in his park or outlet a pleasing incident, he might build one at little or no expense. For whenever he had occasion for a new barn, stable, dog-kennel, or the like structure, it would be only needful to erect this building on the gentle declivity of an hill, with a like rising opposite to it, at a few hundred yards distance; and perhaps success might be the easier ensured could some canal, lake, or stream, intervene. From a seat at the *centrum phonicum* he and his friends might amuse themselves sometimes of an evening with the prattle of this loquacious nymph; of whose complacency and decent reserve more may be said than can with truth of every individual of her sex; since she is

> . . . quae nec reticere loquenti,
> Nec prior ipsa loqui didicit resonabilis echo.
>
> [. . . resounding Echo, who could neither hold
> her peace when others spoke, nor learn to
> speak first herself.]   [Ovid, *Metamorphoses* 3]

I am, &c.

P.S. The classic reader will, I trust, pardon the following lovely quotation, so finely describing echoes, and so poetically accounting for their causes from popular superstition:

> Quae benè quom videas, rationem reddere possis
> Tute tibi atque aliis, quo pacto per loca sola
> Saxa pareis formas verborum ex ordine reddant,
> Palanteis comites quom monteis inter opacos
> Quaerimus, et magnâ dispersos voce ciemus.
> Sex etiam, aut septem loca vidi reddere voces
> Unam quom jaceres: ita colles collibus ipsis
> Verba repulsantes iterabant dicta referre.
> Haec loca capripedes Satyros, Nymphasque tenere
> Finitimi fingunt, et Faunos esse loquuntur;
> Quorum noctivago strepitu, ludoque jocanti

Adfirmant volgo taciturna silentia rumpi,
Chordarumque sonos fieri, dulceisque querelas,
Tibia quas fundit digitis pulsata canentum:
Et genus agricolûm latè sentiscere, quom Pan
Pinea semiferi capitis velamina quassans,
Unco saepe labro calamos percurrit hianteis,
Fistula silvestrem ne cesset fundere musam.

(Lucretius, [*De Natura Rerum*] *Lib.* IV. l. 576)

[When you perceive this well, you may be able to give a reason to yourself
and others, how it is in solitary places that the rocks give back the same
shapes of words in their order, when we seek straying comrades amongst
the shady mountains and call loudly upon them to all sides. I have even
seen places give back six or seven cries, when you uttered one: so did hill
to hill themselves, buffeting back the words, repeat the reverberations.

Such places the neighbours imagine to hold goatfoot satyrs and nymphs,
and they say there are fauns, by whose night-wandering noise and jocund
play they commonly declare the voiceless silence to be broken, with the
sound of strings and sweet plaintive notes, which the pipe sends forth
touched by the player's fingers; they tell how the farmers' men all over the
countryside listen, while Pan, shaking the pine leaves that cover his half-
human head, often runs over the open reeds with curved lips, that the
panpipes may never slacken in their flood of woodland music.]

LETTER 39

Selborne, May 13, 1778

Dear Sir,

Among the many singularities attending those amusing birds
the swifts, I am now confirmed in the opinion that we have every year
the same number of pairs invariably; at least the result of my inquiry
has been exactly the same for a long time past. The swallows and
martins are so numerous, and so widely distributed over the village,
that it is hardly possible to recount them; while the swifts, though
they do not all build in the church, yet so frequently haunt it, and
play and rendezvous round it, that they are easily enumerated. The
number that I constantly find are eight pairs; about half of which
reside in the church, and the rest build in some of the lowest and

meanest thatched cottages. Now as these eight pairs, allowance being
made for accidents, breed yearly eight pairs more, what becomes
annually of this increase; and what determines every spring which
pairs shall visit us, and reoccupy their ancient haunts?

Ever since I have attended to the subject of ornithology, I have
always supposed that that sudden reverse of affection, that strange
ἀντιστοργή which immediately succeeds in the feathered kind to
the most passionate fondness, is the occasion of an equal dispersion of
birds over the face of the earth. Without this provision one favourite
district would be crowded with inhabitants, while others would be
destitute and forsaken. But the parent birds seem to maintain a jeal-
ous superiority, and to oblige the young to seek for new abodes: and
the rivalry of the males, in many kinds, prevents their crowding the
one on the other. Whether the swallows and house-martins return in
the same exact number annually is not easy to say, for reasons given
above: but it is apparent, as I have remarked before in my monographies,* that the numbers returning bear no manner of proportion to
the numbers retiring.

LETTER 40

Selborne, June 2, 1778

Dear Sir,

The standing objection to botany has always been, that it is a pursuit that amuses the fancy and exercises the memory, without
improving the mind or advancing any real knowledge: and, where the
science is carried no farther than a mere systematic classification, the
charge is but too true. But the botanist that is desirous of wiping off
this aspersion should be by no means content with a list of names; he
should study plants philosophically, should investigate the laws of
vegetation, should examine the powers and virtues of efficacious
herbs, should promote their cultivation; and graft the gardener, the
planter, and the husbandman, on the phytologist.* Not that system is
by any means to be thrown aside; without system the field of Nature
would be a pathless wilderness; but system should be subservient to,
not the main object of, pursuit.

Vegetation is highly worthy of our attention; and in itself is of the utmost consequence to mankind, and productive of many of the greatest comforts and elegancies of life. To plants we owe timber, bread, beer, honey, wine, oil, linen, cotton, &c. what not only strengthens our hearts, and exhilerates our spirits, but what secures us from inclemencies of weather and adorns our persons. Man, in his true state of nature, seems to be subsisted by spontaneous vegetation:* in middle climes, where grasses prevail, he mixes some animal food with the produce of the field and garden: and it is towards the polar extremes only that, like his kindred bears and wolves, he gorges himself with flesh alone, and is driven, to what hunger has never been known to compel the very beasts, to prey on his own species.[1]*

The productions of vegetation have had a vast influence on the commerce of nations, and have been the great promoters of navigation, as may be seen in the articles of sugar, tea, tobacco, opium, ginseng, betel, paper, &c. As every climate has its peculiar produce, our natural wants bring on a mutual intercourse; so that by means of trade each distant part is supplied with the growth of every latitude. But, without the knowledge of plants and their culture, we must have been content with our hips and haws, without enjoying the delicate fruits of India and the salutiferous* drugs of Peru.

Instead of examining the minute distinctions of every various species of each obscure genus, the botanist should endeavour to make himself acquainted with those that are useful. You shall see a man readily ascertain every herb of the field, yet hardly know wheat from barley, or at least one sort of wheat or barley from another.

But of all sorts of vegetation the grasses seem to be most neglected; neither the farmer nor the grazier seem to distinguish the annual from the perennial, the hardy from the tender, nor the succulent and nutritive from the dry and juiceless.

The study of grasses would be of great consequence to a northerly, and grazing kingdom. The botanist that could improve the swerd* of the district where he lived would be an useful member of society: to raise a thick turf on a naked soil would be worth volumes of systematic knowledge; and he would be the best commonwealth's man that could occasion the growth of 'two blades of grass where one alone was seen before.'*

I am, &c.

---

[1] See the late Voyages to the south-seas.

LETTER 41

Selborne, July 3, 1778

Dear Sir,

In a district so diversified with such a variety of hill and dale, aspects, and soils, it is no wonder that great choice of plants should be found. Chalks, clays, sands, sheep-walks and downs, bogs, heaths, woodlands, and champaign* fields, cannot but furnish an ample Flora. The deep rocky lanes abound with filices,* and the pastures and moist woods with fungi. If in any branch of botany we may seem to be wanting, it must be in the large aquatic plants, which are not to be expected on a spot far removed from rivers, and lying up amidst the hill country at the spring heads. To enumerate all the plants that have been discovered within our limits would be a needless work; but a short list of the more rare, and the spots where they are to be found, may be neither unacceptable nor unentertaining:—

Helleborus foetidus, stinking hellebore, bear's foot, or setter-wort, all over the High-wood and Coney-croft-hanger: this continues a great branching plant the winter through, blossoming about January, and is very ornamental in shady walks and shrubberies. The good women give the leaves powdered to children troubled with worms; but it is a violent remedy, and ought to be administered with caution.

Helleborus viridis, green hellebore,—in the deep stony lane on the left hand just before the turning to Norton-farm, and at the top of Middle Dorton under the hedge: this plant dies down to the ground early in autumn, and springs again about February, flowering almost as soon as it appears above ground.

Vaccinium oxycoccos, creeping bilberries, or cranberries,—in the bogs of Bin's-pond;

Vaccinium myrtillus, whortle, or bilberries,—on the dry hillocks of Wolmer-forest;

Drosera rotundifolia,   round-leaved sundew.   In the bogs
————*longifolia*,   long-leaved ditto.   of Bin's-pond.

Comarum palustre, purple comarum, or marsh cinque foil,—in the bogs of Bins-pond;

Hypericum androsaemum, Tutsan, St. John's Wort,—in the stony, hollow lanes;

Vinca minor, less periwinkle,—in Selborne-hanger and Shrub-wood;

Monotropa hypopithys, yellow monotropa, or birds' nest,—in Selborne-hanger under the shady beeches, to whose roots it seems to be parasitical—at the north-west end of the Hanger;

*Chlora perfoliata, Blackstonia perfoliata, Hudsoni,** perfoliated yellow-wort,—on the banks in the King's-field;

Paris quadrifolia, herb Paris, true-love, or one-berry,—in the Church-litten-coppice;

Chrysosplenium oppositifolium, opposite golden saxifrage,—in the dark and rocky hollow lanes;

Gentiana amarella, autumnal gentian, or fellwort,—on the Zigzag and Hanger;

Lathraea squammaria, tooth-wort,—in the Church-litten-coppice under some hazels near the foot-bridge, in Trimming's garden hedge, and on the dry wall opposite Grange-yard;

Dipsacus pilosus, small teasel,—in the Short and Long Lith.

Lathyrus sylvestris, narrow-leaved, or wild lathyrus,—in the bushes at the foot of the Short Lith, near the path;

Ophrys spiralis, ladies traces,—in the Long Lith, and towards the south-corner of the common;

Ophrys nidus avis, birds' nest ophrys,—in the Long Lith under the shady beeches among the dead leaves; in Great Dorton among the bushes, and on the Hanger plentifully;

Serapias latifolia, helleborine,—in the High-wood under the shady beeches;

Daphne laureola, spurge laurel,—in Selborne-Hanger and the High-wood;

Daphne mezereum, the mezereon,—in Selborne-Hanger among the shrubs at the south-east end above the cottages.

Lycoperdon tuber, truffles,—in the Hanger and High-wood.

Sambucus ebulus,—dwarf elder, walwort, or danewort,—among the rubbish and ruined foundations of the Priory.

Of all the propensities of plants none seem more strange than their different periods of blossoming. Some produce their flowers in the winter, or very first dawnings of spring; many when the spring is established; some at midsummer, and some not till autumn. When we see the *helleborus foetidus* and *helleborus niger* blowing* at Christmas, the *helleborus hyemalis* in January, and the *helleborus* *viridis* as soon as ever it emerges out of the ground, we do not wonder, because they are kindred plants that we expect should keep pace the one with the other. But other congenerous vegetables differ so widely in their time of flowering, that we cannot but admire. I shall only instance at present in the *crocus sativus*, the vernal, and the autumnal crocus, which have such an affinity, that the best botanists only make them varieties of the same genus, of which there is only one species; not being able to discern any difference in the corolla,* or in the internal structure. Yet the vernal crocus expands its flowers by the beginning of March at farthest, and often in very rigorous weather; and cannot be retarded but by some violence offered:—while the autumnal (the Saffron) defies the influence of the spring and summer, and will not blow till most plants begin to fade and run to seed. This circumstance is one of the wonders of the creation, little noticed, because a common occurrence; yet ought not to be overlooked on account of its being familiar, since it would be as difficult to be explained as the most stupendous phaenomenon in nature.

> Say, what impels, amidst surrounding snow
> Congeal'd, the crocus' flamy bud to glow?
> Say, what retards, amidst the summer's blaze,
> Th' autumnal bulb, till pale, declining days?
> The GOD of SEASONS; whose pervading power
> Controls the sun, or sheds the fleecy shower:
> He bids each flower his quick'ning word obey;
> Or to each lingering bloom enjoins delay.*

LETTER 42

> Omnibus animalibus reliquis certus et uniusmodi, et
> in suo cuique genere incessus est: aves solae vario
> meatu feruntur, et in terrâ, et in äere.

> [All other creatures, in their particular classes, have a
> distinctive mode of locomotion; only birds move by
> different means both on land and in the air.]

> [Pliny, *Natural History*, 10. 38]

Selborne, Aug. 7, 1778

Dear Sir,

A good ornithologist should be able to distinguish birds by their air* as well as by their colours and shape; on the ground as well as on the wing, and in the bush as well as in the hand. For, though it must not be said that every species of birds has a manner peculiar to itself, yet there is somewhat in most genera at least, that at first sight discriminates them, and enables a judicious observer to pronounce upon them with some certainty. Put a bird in motion

. . . Et vera incessu patuit . . .

[And the true [bird] is clear from its gait.]

[Virgil, *Aeneid* I]

Thus kites and buzzards sail round in circles with wings expanded and motionless; and it is from their gliding manner that the former are still called in the north of England gleads, from the Saxon verb *glidan*, to glide. The kestrel, or wind-hover, has a peculiar mode of hanging in the air in one place, his wings all the while being briskly agitated. Hen-harriers fly low over heaths or fields of corn, and beat the ground regularly like a pointer or setting-dog. Owls move in a buoyant manner, as if lighter than the air; they seem to want ballast. There is a peculiarity belonging to ravens that must draw the attention even of the most incurious—they spend all their leisure time in striking and cuffing each other on the wing in a kind of playful skirmish; and, when they move from one place to another, frequently turn on their backs with a loud croak, and seem to be falling to the ground. When this odd gesture betides them, they are scratching themselves with one foot, and thus lose the center of gravity. Rooks sometimes dive and tumble in a frolicksome manner; crows and daws swagger in their walk; wood-peckers, fly *volatu undoso* [in an undulating fashion], opening and closing their wings at every stroke, and so are always rising or falling in curves. All of this genus use their tails, which incline downward, as a support while they run up trees. Parrots, like all other hooked-clawed birds, walk aukwardly, and make use of their bill as a third foot, climbing and descending with ridiculous caution. All the gallinae parade and walk gracefully, and run nimbly; but fly with difficulty, with an impetuous whirring, and in a straight line. Magpies and jays stutter with powerless wings,

LITTLE AUK.

and make no dispatch; herons seem incumbered with too much sail for their light bodies; but these vast hollow wings are necessary in carrying burdens, such as large fishes, and the like; pigeons, and particularly the sort called smiters, have a way of clashing their wings the one against the other over their backs with a loud snap; another variety called tumblers turn themselves over in the air. Some birds have movements peculiar to the season of love: thus ring-doves, though strong and rapid at other times, yet in the spring hang about on the wing in a toying and playful manner; thus the cock-snipe, while breeding, forgetting his former flight, fans the air like the windhover; and the green-finch in particular exhibits such languishing and faultering gestures as to appear like a wounded and dying bird; the king-fisher darts along like an arrow; fern-owls, or goat-suckers, glance in the dusk over the tops of trees like a meteor; starlings as it were swim along, while missel-thrushes use a wild and desultory flight; swallows sweep over the surface of the ground and water, and

COOT.

distinguish themselves by rapid turns and quick evolutions; swifts dash round in circles; and the bank-martin moves with frequent vacillations like a butterfly. Most of the small birds fly by jerks, rising and falling as they advance. Most small birds hop; but wagtails and larks walk, moving their legs alternately. Skylarks rife and fall perpendicularly as they sing; woodlarks hang poised in the air; and titlarks rise and fall in large curves, singing in their descent. The white-throat uses odd jerks and gesticulations over the tops of hedges and bushes. All the duck-kind waddle; divers and auks walk as if fettered, and stand erect on their tails: these are the compedes of Linnaeus. Geese and cranes, and most wild-fowls, move in figured flights, often changing their position. The secondary remiges* of *Tringae*, wild-ducks, and some others, are very long, and give their wings, when in motion, an hooked appearance. Dabchicks, moorhens, and coots, fly erect, with their legs hanging down, and hardly make any dispatch; the reason is plain, their wings are placed too forward out of the true center of gravity; as the legs of auks and divers are situated too backward.

LETTER 43

Selborne, Sept. 9, 1778

Dear Sir,

From the motion of birds, the transition is natural enough to their notes and language, of which I shall say something. Not that I would pretend to understand their language like the vizier; who, by the recital of a conversation which passed between two owls, reclaimed a sultan,[1] before delighting in conquest and devastation; but I would be thought only to mean that many of the winged tribes have various sounds and voices adapted to express their various passions, wants, and feelings; such as anger, fear, love, hatred, hunger, and the like. All species are not equally eloquent; some are copious and fluent as it were in their utterance, while others are confined to a few important sounds: no bird, like the fish kind, is quite mute, though some are rather silent. The language of birds is very ancient, and, like other ancient modes of speech, very elliptical; little is said, but much is meant and understood.

The notes of the eagle-kind are shrill and piercing; and about the season of nidification much diversified, as I have been often assured by a curious observer of Nature, who long resided at Gibraltar, where eagles abound. The notes of our hawks much resemble those of the king of birds. Owls have very expressive notes; they hoot in a fine vocal sound, much resembling the *vox humana* [human voice], and reducible by a pitch-pipe to a musical key. This note seems to express complacency and rivalry among the males: they use also a quick call and an horrible scream; and can snore and hiss when they mean to menace. Ravens, besides their loud croak, can exert a deep and solemn note that makes the woods to echo; the amorous sound of a crow is strange and ridiculous; rooks, in the breeding season, attempt sometimes in the gaiety of their hearts to sing, but with no great success; the parrot-kind have many modulations of voice, as appears by their aptitude to learn human sounds; doves coo in an amorous and mournful manner, and are emblems of despairing lovers; the woodpecker sets up a sort of loud and hearty laugh; the fern-owl, or goat-sucker, from the dusk till day-break, serenades

---

[1] See *Spectator*, Vol. VII, No. 512.*

his mate with the clattering of castanets. All the tuneful passeres express their complacency by sweet modulations, and a variety of melody. The swallow, as has been observed in a former letter, by a shrill alarm bespeaks the attention of the other hirundines, and bids them be aware that the hawk is at hand. Aquatic and gregarious birds, especially the nocturnal, that shift their quarters in the dark, are very noisy and loquacious; as cranes, wild-geese, wild-ducks, and the like: their perpetual clamour prevents them from dispersing and losing their companions.

In so extensive a subject, sketches and outlines are as much as can be expected; for it would be endless to instance in all the infinite variety of the feathered nation. We shall therefore confine the remainder of this letter to the few domestic fowls of our yards, which are most known, and therefore best understood. And first the peacock, with his gorgeous train, demands our attention; but, like most of the gaudy birds, his notes are grating and shocking to the ear: the yelling of cats, and the braying of an ass, are not more disgustful. The voice of the goose is trumpet-like, and clanking; and once saved the Capitol at Rome, as grave historians assert:* the hiss also of the gander is formidable and full of menace, and 'protective of his young.' Among ducks the sexual distinction of voice is remarkable; for, while the quack of the female is loud and sonorous, the voice of the drake is inward and harsh, and feeble, and scarce discernible. The cock turkey struts and gobbles to his mistress in a most uncouth manner; he hath also a pert and petulant note when he attacks his adversary. When a hen turkey leads forth her young brood she keeps a watchful eye; and if a bird of prey appear, though ever so high in the air, the careful mother announces the enemy with a little inward moan, and watches him with a steady and attentive look; but, if he approach, her note becomes earnest and alarming, and her outcries are redoubled.

No inhabitants of a yard seem possessed of such a variety of expression and so copious a language as common poultry. Take a chicken of four or five days old, and hold it up to a window where there are flies, and it will immediately seize its prey, with little twitterings of complacency; but if you tender it a wasp or a bee, at once its note becomes harsh, and expressive of disapprobation and a sense of danger. When a pullet is ready to lay she intimates the event by a joyous and easy soft note. Of all the occurrences of their life that of laying seems to be the most important; for no sooner has a hen disburdened herself, than she rushes forth with a clamorous kind of joy, which the cock and the rest of his mistresses immediately adopt. The tumult is not confined to the family concerned, but catches from yard to yard, and spreads to every homestead within hearing, till at last the whole village is in an uproar. As soon as a hen becomes a mother her new relation demands a new language; she then runs clocking and screaming about, and seems agitated as if possessed. The father of the flock has also a considerable vocabulary; if he finds food, he calls a favourite concubine to partake; and if a bird of prey passes over, with a warning voice he bids his family beware. The gallant chanticleer* has, at command, his amorous phrases and his terms of defiance. But the sound by which he is best known is his crowing: by this he has been distinguished in all ages as the countryman's clock or larum, as the watchman that proclaims the divisions of the night. Thus the poet elegantly styles him:

> . . . the crested cock, whose clarion sound
> The silent hours.
> [Milton, *Paradise Lost*, VII. 443–4]

A neighbouring gentleman one summer had lost most of his chickens by a sparrow-hawk, that came gliding down between a faggot pile and the end of his house to the place where the coops stood. The owner, inwardly vexed to see his flock thus diminishing, hung a setting net adroitly between the pile and the house, into which the caitif* dashed, and was entangled. Resentment suggested the law of retaliation; he therefore clipped the hawk's wings, cut off his talons, and, fixing a cork on his bill, threw him down among the brood-hens. Imagination cannot paint the scene that ensued; the expressions that fear, rage, and revenge, inspired, were new, or at least such as had been unnoticed before: the exasperated matrons upbraided, they execrated, they insulted, they triumphed. In a word, they never desisted from buffetting their adversary till they had torn him in an hundred pieces.

LETTER 44

Selborne

. . . monstrent

. . . .

Quid tantùm Oceano properent se tingere soles
Hyberni; vel quae tardis mora noctibus obstet.

[Let (the Muses) show . . . why the winter sun hastens so fast
(across the sky) only to dip into the ocean; or, what impedi-
ments hinder the slow-moving nights.]   [Virgil, *Georgics* 2]

Gentlemen who have outlets* might contrive to make ornament subservient to utility: a pleasing eye-trap* might also contribute to promote science: an obelisk in a garden or park might be both an embellishment and an heliotrope.*

Any person that is curious, and enjoys the advantage of a good horizon, might, with little trouble, make two heliotropes; the one for the winter, the other for the summer solstice: and these two erections might be constructed with very little expense; for two pieces of timber frame-work, about ten or twelve feet high, and four feet broad at the base, and close lined with plank, would answer the purpose.

The erection for the former should, if possible, be placed within sight of some window in the common sitting parlour; because men, at that dead season of the year, are usually within doors at the close of the day; while that for the latter might be fixed for any given spot in the garden or outlet: whence the owner might contemplate, in a fine summer's evening, the utmost extent that the sun makes to the north-ward at the season of the longest days. Now nothing would be neces-sary but to place these two objects with so much exactness, that the westerly limb of the sun, at setting, might but just clear the winter heliotrope to the west of it on the shortest day; and that the whole disc of the sun, at the longest day, might exactly at setting also clear the summer heliotrope to the north of it.

By this simple expedient it would soon appear that there is no such thing, strictly speaking, as a solstice; for, from the shortest day, the owner would, every clear evening, see the disc advancing, at its set-ting, to the westward of the object; and, from the longest day, observe the sun retiring backwards every evening at its setting, towards the object westward, till, in a few nights, it would set quite behind it, and so by degrees to the west of it: for when the sun comes near the summer solstice, the whole disc of it would at first set behind the object; after a time the northern limb would first appear, and so every night gradually more, till at length the whole diameter would set northward of it for about three nights; but on the middle night of the three, sensibly more remote than the former or following. When beginning its recess from the summer tropic, it would continue more and more to be hidden every night, till at length it would descend quite behind the object again; and so nightly more and more to the westward.

LETTER 45

Selborne

. . . Mugire videbis
Sub pedibus terram, et descendere montibus ornos.

[You will experience the earth rumbling beneath your
feet and the ash trees falling down the mountainsides.]

[Virgil, *Aeneid* 4]

When I was a boy I used to read, with astonishment and implicit assent, accounts in Baker's *Chronicle* of walking hills and travelling mountains.* John Philips, in his *Cyder*, alludes to the credit that was given to such stories with a delicate but quaint* vein of humour peculiar to the author of the *Splendid Shilling*.

> I nor advise, nor reprehend the choice
> Of Marcley Hill; the apple no where finds
> A kinder mould; yet 'tis unsafe to trust
> Deceitful ground: who knows but that once more
> This mount may journey, and his present site
> Forsaken, to thy neighbour's bounds transfer
> Thy goodly plants, affording matter strange
> For law debates!

But, when I came to consider better, I began to suspect that though our hills may never have journeyed far, yet that the ends of many of them have slipped and fallen away at distant periods, leaving the cliffs bare and abrupt. This seems to have been the case with Nore and Whetham Hills; and especially with the ridge between Harteley Park and Word le ham, where the ground has slid into vast swellings and furrows; and lies still in such romantic confusion as cannot be accounted for from any other cause. A strange event, that happened not long since, justifies our suspicions; which, though it befell not within the limits of this parish, yet as it was within the hundred* of Selborne, and as the circumstances were singular, may fairly claim a place in a work of this nature.

The months of January and February, in the year 1774, were remarkable for great melting snows and vast gluts of rain; so that by the end of the latter month the land-springs, or lavants, began to prevail, and to be near as high as in the memorable winter of 1764.* The beginning of March also went on in the same tenor; when, in the night between the 8th and 9th of that month, a considerable part of the great woody hanger at Hawkley was torn from its place, and fell down, leaving a high free-stone cliff naked and bare, and resembling the steep side of a chalk-pit. It appears that this huge fragment, being perhaps sapped and undermined by waters, foundered, and was ingulfed, going down in a perpendicular direction; for a gate which stood in the field, on the top of the hill, after sinking with its posts for thirty or forty feet, remained in so true and upright a position as to open and shut with great exactness, just as in its first situation. Several oaks also are still standing, and in a state of vegetation, after taking the same desperate leap.* That great part of this prodigious mass was absorbed in some gulf below, is plain also from the inclining ground at the bottom of the hill, which is free and unincumbered; but would have been buried in heaps of rubbish, had the fragment parted and fallen forward. About an hundred yards from the foot of this hanging coppice stood a cottage by the side of a lane; and two hundred yards lower, on the other side of the lane, was a farm-house, in which lived a labourer and his family; and, just by, a stout new barn. The cottage was inhabited by an old woman, and her son, and his wife. These people in the evening, which was very dark and tempestuous, observed that the brick floors of their kitchens began to heave and part; and that the walls seemed to open, and the roofs to crack: but they all agree that no tremor of the ground, indicating an earthquake, was ever felt; only that the wind continued to make a most

tremendous roaring in the woods and hangers. The miserable inhabitants, not daring to go to bed, remained in the utmost solicitude and
confusion, expecting every moment to be buried under the ruins of
their shattered edifices. When day-light came they were at leisure to
contemplate the devastations of the night: they then found that a
deep rift, or chasm, had opened under their houses, and torn them, as
it were, in two; and that one end of the barn had suffered in a similar
manner; that a pond near the cottage had undergone a strange reverse,
becoming deep at the shallow end, and so vice versa; that many large
oaks were removed out of their perpendicular, some thrown down,
and some fallen into the heads of neighbouring trees; and that a gate
was thrust forward, with its hedge, full six feet, so as to require a new
track to be made to it. From the foot of the cliff the general course of
the ground, which is pasture, inclines in a moderate descent for half
a mile, and is interspersed with some hillocks, which were rifted, in
every direction, as well towards the great woody hanger, as from it. In
the first pasture the deep clefts began; and running across the lane,
and under the buildings, made such vast shelves that the road was
impassable for some time; and so over to an arable field on the other
side, which was strangely torn and disordered. The second pasture
field, being more soft and springy, was protruded forward without
many fissures in the turf, which was raised in long ridges resembling
graves, lying at right angles to the motion. At the bottom of this
enclosure the soil and turf rose many feet against the bodies of some
oaks that obstructed their farther course and terminated this awful
commotion.

The perpendicular height of the precipice, in general, is twenty-
three yards; the length of the lapse, or slip, as seen from the fields
below, one hundred and eighty-one; and a partial fall, concealed in
the coppice, extends seventy yards more: so that the total length of
this fragment that fell was two hundred and fifty-one yards. About
fifty acres of land suffered from this violent convulsion; two houses
were entirely destroyed; one end of a new barn was left in ruins, the
walls being cracked through the very stones that composed them; a
hanging coppice was changed to a naked rock; and some grass grounds
and an arable field so broken and rifted by the chasms as to be rendered, for a time, neither fit for the plough or safe for pasturage, till
considerable labour and expense had been bestowed in levelling the
surface and filling in the gaping fissures.

LETTER 46

Selborne

. . . resonant arbusta . . .

[the woods resound (with the shrill voice of
the grasshopper).]   [Virgil, *Eclogues* 2]

There is a steep abrupt pasture field interspersed with furze* close to
the back of this village, well known by the name of the Short Lithe,
consisting of a rocky dry soil, and inclining to the afternoon sun. This
spot abounds with the *gryllus campestris*, or field-cricket; which,
though frequent in these parts, is by no means a common insect in
many other counties.

As their cheerful summer cry cannot but draw the attention of a
naturalist, I have often gone down to examine the oeconomy of these
grylli, and study their mode of life: but they are so shy and cautious
that it is no easy matter to get a sight of them; for, feeling a person's
footsteps as he advances, they stop short in the midst of their song,
and retire backward nimbly into their burrows, where they lurk till all
suspicion of danger is over.

At first we attempted to dig them out with a spade, but without any
great success; for either we could not get to the bottom of the hole,
which often terminated under a great stone; or else, in breaking up
the ground, we inadvertently squeezed the poor insect to death. Out
of one so bruised we took a multitude of eggs, which were long and
narrow, of a yellow colour, and covered with a very tough skin. By
this accident we learned to distinguish the male from the female; the
former of which is shining black, with a golden stripe across his
shoulders; the latter is more dusky, more capacious about the abdomen, and carries a long sword-shaped weapon at her tail, which probably is the instrument with which she deposits her eggs in crannies
and safe receptacles.

Where violent methods will not avail, more gentle means will often
succeed; and so it proved in the present case; for, though a spade be
too boisterous and rough an implement, a pliant stalk of grass, gently
insinuated into the caverns, will probe their windings to the bottom,
and quickly bring out the inhabitant; and thus the humane inquirer

may gratify his curiosity without injuring the object of it. It is remarkable that, though these insects are furnished with long legs behind, and brawny thighs for leaping, like grasshoppers; yet when driven from their holes they shew no activity, but crawl along in a shiftless manner, so as easily to be taken: and again, though provided with a curious apparatus of wings, yet they never exert them when there seems to be the greatest occasion. The males only make that shrilling noise perhaps out of rivalry and emulation, as is the case with many animals which exert some sprightly note during their breeding time: it is raised by a brisk friction of one wing against the other. They are solitary beings, living singly male or female, each as it may happen; but there must be a time when the sexes have some intercourse, and then the wings may be useful perhaps during the hours of night. When the males meet they will fight fiercely, as I found by some which I put into the crevices of a dry stone wall, where I should have been glad to have made them settle. For though they seemed distressed by being taken out of their knowledge,* yet the first that got possession of the chinks would seize on any that were obtruded upon them with a vast row of serrated fangs. With their strong jaws, toothed like the shears of a lobster's claws, they perforate and round their curious regular cells, having no fore-claws to dig, like the mole-cricket. When taken in hand I could not but wonder that they never offered to defend themselves, though armed with such formidable weapons. Of such herbs as grow before the mouths of their burrows they eat indiscriminately; and on a little platform, which they make just by, they drop their dung; and never, in the day time, seem to stir more than two or three inches from home. Sitting in the entrance of their caverns they chirp all night as well as day from the middle of the month of May to the middle of July; and in hot weather, when they are most vigorous, they make the hills echo; and, in the stiller hours of darkness, may be heard to a considerable distance. In the beginning of the season their notes are more faint and inward; but become louder as the summer advances, and so die away again by degrees.

Sounds do not always give us pleasure according to their sweetness and melody; nor do harsh sounds always displease. We are more apt to be captivated or disgusted with the associations which they promote, than with the notes themselves. Thus the shrilling of the field-cricket, though sharp and stridulous, yet marvellously delights some hearers, filling their minds with a train of summer ideas of every thing that is rural, verdurous, and joyous.

About the tenth of March the crickets appear at the mouths of their cells, which they then open and bore, and shape very elegantly. All that ever I have seen at that season were in their pupa state, and had only the rudiments of wings, lying under a skin or coat, which must be cast before the insect can arrive at its perfect state;[1] from whence I should suppose that the old ones of last year do not always survive the winter. In August their holes begin to be obliterated, and the insects are seen no more till spring.

Not many summers ago I endeavoured to transplant a colony to the terrace in my garden, by boring deep holes in the sloping turf.* The new inhabitants stayed some time, and fed and sung; but wandered away by degrees, and were heard at a farther distance every morning; so that it appears that on this emergency they made use of their wings in attempting to return to the spot from which they were taken.

One of these crickets, when confined in a paper cage and set in the sun, and supplied with plants moistened with water, will feed and thrive, and become so merry and loud as to be irksome in the same room where a person is sitting: if the plants are not wetted it will die.

LETTER 47

Selborne

Dear Sir,

> Far from all resort of mirth
> Save the cricket on the hearth.
>
> Milton's 'Il Penseroso' [lines 81–2].

While many other insects must be sought after in fields and woods, and waters, the *gryllus domesticus*, or house-cricket, resides altogether within our dwellings, intruding itself upon our notice whether we will or no. This species delights in new-built houses, being, like the spider, pleased with the moisture of the walls; and besides, the softness of the

---

[1] We have observed that they cast these skins in April, which are then seen lying at the mouths of their holes.

mortar enables them to burrow and mine between the joints of the bricks or stones, and to open communications from one room to another. They are particularly fond of kitchens and bakers' ovens, on account of their perpetual warmth.

Tender insects that live abroad either enjoy only the short period of one summer, or else doze away the cold uncomfortable months in profound slumbers; but these, residing as it were in a torrid zone, are always alert and merry: a good Christmas fire is to them like the heats of the dog-days.* Though they are frequently heard by day, yet is their natural time of motion only in the night. As soon as it grows dusk, the chirping increases, and they come running forth, and are from the size of a flea to that of their full stature. As one should suppose, from the burning atmosphere which they inhabit, they are a thirsty race, and shew a great propensity for liquids, being found frequently drowned in pans of water, milk, broth, or the like. Whatever is moist they affect; and therefore often gnaw holes in wet woollen stockings and aprons that are hung to the fire: they are the housewife's barometer, foretelling her when it will rain; and are prognostic sometimes, she thinks, of ill or good luck; of the death of a near relation, or the approach of an absent lover. By being the constant companions of her solitary hours they naturally become the objects of her superstition. These crickets are not only very thirsty, but very voracious; for they will eat the scummings of pots, and yeast, salt, and crumbs of bread; and any kitchen offal or sweepings. In the summer we have observed them to fly, when it became dusk, out of the windows, and over the neighbouring roofs. This feat of activity accounts for the sudden manner in which they often leave their haunts, as it does for the method by which they come to houses where they were not known before. It is remarkable, that many sorts of insects seem never to use their wings but when they have a mind to shift their quarters and settle new colonies. When in the air they move '*volatu undoso*,' in waves or curves, like wood-peckers, opening and shutting their wings at every stroke, and so are always rising or sinking.

When they increase to a great degree, as they did once in the house where I am now writing,* they become noisome pests, flying into the candles, and dashing into people's faces; but may be blasted and destroyed by gunpowder discharged into their crevices and crannies. In families, at such times, they are, like Pharaoh's plague of frogs,—'in their bedchambers, and upon their beds, and in their ovens, and in their kneading-troughs.'[1] Their shrilling noise is occasioned by a brisk attrition of their wings. Cats catch hearth-crickets, and, playing with them as they do with mice, devour them. Crickets may be destroyed, like wasps, by phials half filled with beer, or any liquid, and set in their haunts; for, being always eager to drink, they will crowd in till the bottles are full.

LETTER 48

Selborne

How diversified are the modes of life not only of incongruous but even of congenerous animals; and yet their specific distinctions are not more various than their propensities. Thus, while the field-cricket delights in sunny dry banks, and the house-cricket rejoices amidst the glowing heat of the kitchen hearth or oven, the *gryllus gryllo talpa* (the mole-cricket), haunts moist meadows, and frequents the sides of ponds and banks of streams, performing all its functions in a swampy wet soil. With a pair of fore-feet, curiously adapted to the purpose, it burrows and works under ground like the mole, raising a ridge as it proceeds, but seldom throwing up hillocks.

As mole-crickets often infest gardens by the sides of canals, they are unwelcome guests to the gardener, raising up ridges in their subterraneous progress, and rendering the walks unsightly. If they take to the kitchen quarters, they occasion great damage among the plants and roots, by destroying whole beds of cabbages, young legumes, and flowers. When dug out they seem very slow and helpless, and make no use of their wings by day; but at night they come abroad, and make long excursions, as I have been convinced by finding stragglers, in a morning, in improbable places. In fine weather, about the middle of April, and just at the close of day, they begin to solace themselves with a low, dull, jarring note, continued for a long time without interruption, and not unlike the chattering of the fern-owl, or goat-sucker, but more inward.

About the beginning of May they lay their eggs, as I was once an eye-witness: for a gardener at an house, where I was on a visit,*

---

[1] Exodus 8. 3.

happening to be mowing, on the 6th of that month, by the side of a canal, his scythe struck too deep, pared off a large piece of turf, and laid open to view a curious scene of domestic oeconomy:

> ... ingentem lato dedit ore fenestram:
> Apparet domus intus, et atria longa patescunt:
> Apparent ... penetralia.

> [(Pyrrhus) made a huge, cavernous opening, and
> the interior of the palace appeared, with its long
> halls clearly visible; beyond, there were the inner
> chambers (of Priam).]   [Virgil, *Aeneid* 2]

There were many caverns and winding passages leading to a kind of chamber, neatly smoothed and rounded, and about the size of a moderate snuff-box. Within this secret nursery were deposited near an hundred eggs of a dirty yellow colour, and enveloped in a tough skin, but too lately excluded to contain any rudiments of young, being full of a viscous substance. The eggs lay but shallow, and within the influence of the sun, just under a little heap of fresh-moved mould, like that which is raised by ants.

When mole-crickets fly they move '*cursu undoso*,' rising and falling in curves, like the other species mentioned before. In different parts of this kingdom people call them fen-crickets, churr-worms, and eve-churrs, all very apposite names.

Anatomists, who have examined the intestines of these insects, astonish me with their accounts; for they say that, from the structure, position, and number of their stomachs, or maws, there seems to be good reason to suppose that this and the two former species ruminate or chew the cud like many quadrupeds!

LETTER 49

Selborne, May 7, 1779

It is now more than forty years that I have paid some attention to the ornithology of this district, without being able to exhaust the subject: new occurrences still arise as long as any inquiries are kept alive.

In the last week of last month five of those most rare birds, too

uncommon to have obtained an English name, but known to naturalists by the terms of *himantopus*, or *loripes*, and *charadrius himantopus*,
were shot upon the verge of Frinsham-pond, a large lake belonging
to the Bishop of Winchester, and lying between Wolmer-forest, and
the town of Farnham, in the county of Surrey. The pond keeper
says there were three brace in the flock; but that, after he had satisfied
his curiosity, he suffered the sixth to remain unmolested. One of
these specimens I procured, and found the length of the legs to be so
extraordinary, that, at first sight, one might have supposed the shanks
had been fastened on to impose on the credulity of the beholder: they
were legs in caricatura; and had we seen such proportions on a
Chinese or Japan screen we should have made large allowances for
the fancy of the draughtsman. These birds are of the plover family,
and might with propriety be called the stilt plovers. Brisson, under
that idea, gives them the apposite name of *l'echasse.** My specimen,
when drawn and stuffed with pepper,* weighed only four ounces
and a quarter, though the naked part of the thigh measured three
inches and an half, and the legs four inches and an half. Hence we
may safely assert that these birds exhibit, weight for inches, incompar-
ably the greatest length of legs of any known bird.* The flamingo, for
instance, is one of the most long legged birds,* and yet it bears no
manner of proportion to the *himantopus*; for a cock flamingo weighs,
at an average, about four pounds avoirdupois; and his legs and thighs
measure usually about twenty inches. But four pounds are fifteen
times and a fraction more than four ounces, and one quarter; and if
four ounces and a quarter have eight inches of legs, four pounds must
have one hundred and twenty inches and a fraction of legs; *viz.* somewhat more than ten feet; such a monstrous proportion as the world
never saw! If you should try the experiment in still larger birds the
disparity would still increase. It must be matter of great curiosity to
see the stilt plover move; to observe how it can wield such a length of
lever with such feeble muscles as the thighs seem to be furnished
with. At best one should expect it to be but a bad walker: but what
adds to the wonder is, that it has no back toe. Now without that steady
prop to support its steps it must be liable, in speculation, to perpetual
vacillations, and seldom able to preserve the true center of gravity.

The old name of *himantopus* is taken from Pliny; and, by an
aukward metaphor, implies that the legs are as slender and pliant as if
cut out of a thong of leather. Neither Willughby nor Ray, in all their

curious researches, either at home or abroad, ever saw this bird. Mr Pennant never met with it in all Great-Britain, but observed it often in the cabinets of the curious at Paris. Hasselquist says that it migrates to Egypt in the autumn: and a most accurate observer of Nature has assured me that he has found it on the banks of the streams in Andalusia.

Our writers record it to have been found only twice in Great-Britain. From all these relations it plainly appears that these long legged plovers are birds of South Europe, and rarely visit our island; and when they do are wanderers and stragglers, and impelled to make so distant and northern an excursion from motives or accidents for which we are not able to account. One thing may fairly be deduced, that these birds come over to us from the continent, since nobody can suppose that a species not noticed once in an age, and of such a remarkable make, can constantly breed unobserved in this kingdom.

LETTER 50

Selborne, April 21, 1780

Dear Sir,

The old Sussex tortoise, that I have mentioned to you so often, is become my property.* I dug it out of its winter dormitory in March last, when it was enough awakened to express its resentments by hissing; and, packing it in a box with earth, carried it eighty miles in post-chaises. The rattle and hurry of the journey so perfectly roused it that, when I turned it out on a border, it walked twice down to the bottom of my garden; however, in the evening, the weather being cold, it buried itself in the loose mould, and continues still concealed.

As it will be under my eye, I shall now have an opportunity of enlarging my observations on its mode of life, and propensities; and perceive already that, towards the time of coming forth, it opens a breathing place in the ground near its head, requiring, I conclude, a freer respiration as it becomes more alive. This creature not only goes under the earth from the middle of November to the middle of April, but sleeps great part of the summer; for it goes to bed in the longest days at four in the afternoon, and often does not stir in the morning till late. Besides, it retires to rest for every shower; and does not move at all in wet days.

When one reflects on the state of this strange being, it is a matter of wonder to find that Providence should bestow such a profusion of days, such a seeming waste of longevity, on a reptile that appears to relish it so little as to squander more than two thirds of its existence in a joyless stupor, and be lost to all sensation for months together in the profoundest of slumbers.

While I was writing this letter, a moist and warm afternoon, with the thermometer at 50, brought forth troops of shell-snails; and, at the same juncture, the tortoise heaved up the mould and put out its head; and the next morning came forth, as it were raised from the dead; and walked about till four in the afternoon. This was a curious coincidence! a very amusing occurrence! to see such a similarity of feelings between the two φερέοικοι [house-carriers]! for so the Greeks called both the shell-snail and the tortoise.

Summer birds are, this cold and backward spring, unusually late: I have seen but one swallow yet. This conformity with the weather convinces me more and more that they sleep in the winter.

More PARTICULARS respecting the OLD FAMILY TORTOISE, omitted in the Natural History.*

Because we call this creature an abject reptile, we are too apt to under-value his abilities, and depreciate his powers of instinct. Yet he is, as Mr Pope says of his lord,

> . . . Much too wise to walk into a well:*

and has so much discernment as not to fall down an haha; but to stop and withdraw from the brink with the readiest precaution.*

Though he loves warm weather he avoids the hot sun; because his thick shell, when once heated, would, as the poet says of solid armour—'scald with safety.'* He therefore spends the more sultry hours under the umbrella of a large cabbage-leaf, or amidst the waving forests of an asparagus-bed.

But as he avoids heat in the summer, so, in the decline of the year, he improves the faint autumnal beams, by getting within the reflection of a fruit-wall: and, though he never has read that planes inclining to the horizon receive a greater share of warmth,[1] he inclines his shell, by tilting it against the wall, to collect and admit every feeble ray.

Pitiable seems the condition of this poor embarrassed reptile: to be cased in a suit of ponderous armour, which he cannot lay aside; to be imprisoned, as it were, within his own shell, must preclude, we should suppose, all activity and disposition for enterprize. Yet there is a season of the year (usually the beginning of June) when his exertions are remarkable. He then walks on tiptoe, and is stirring by five in the morning; and, traversing the garden, examines every wicket and interstice in the fences, through which he will escape if possible: and often has eluded the care of the gardener, and wandered to some distant field. The motives that impel him to undertake these rambles seem to be of the amorous kind: his fancy then becomes intent on sexual attachments, which transport him beyond his usual gravity, and induce him to forget for a time his ordinary solemn deportment.

LETTER 51

Selborne, Sept. 3, 1781

I have now read your miscellanies through with much care and satisfaction; and am to return you my best thanks for the honourable mention made in them of me as a naturalist, which I wish I may deserve.*

In some former letters I expressed my suspicions that many of the house-martins do not depart in the winter far from this village. I therefore determined to make some search about the south-east end of the hill, where I imagined they might slumber out the uncomfortable months of winter. But supposing that the examination would be made to the best advantage in the spring, and observing that no martins had appeared by the 11th of April last; on that day I employed some men to explore the shrubs and cavities of the suspected spot. The persons took pains, but without any success; however, a remarkable incident occurred in the midst of our pursuit—while the labourers were at work a house-martin, the first that had been seen this year,

[1] Several years ago a book was written entitled *Fruit-walls improved by inclining them to the horizon*: in which the author has shewn, by calculation, that a much greater number of the rays of the sun will fall on such walls than on those which are perpendicular.*

came down the village in the sight of several people, and went at once into a nest, where it stayed a short time, and then flew over the houses; for some days after no martins were observed, not till the 16th of April, and then only a pair. Martins in general were remarkably late this year.

LETTER 52

Selborne, Sept. 9, 1781

I have just met with a circumstance respecting swifts, which furnishes an exception to the whole tenor of my observations ever since I have bestowed any attention on that species of hirundines. Our swifts, in general, withdrew this year about the first day of August, all save one pair, which in two or three days was reduced to a single bird. The perseverance of this individual made me suspect that the strongest of motives, that of an attachment to her young, could alone occasion so late a stay. I watched therefore till the twenty-fourth of August, and then discovered that, under the eaves of the church, she attended upon two young, which were fledged, and now put out their white chins from a crevice. These remained till the twenty-seventh, looking more alert every day, and seeming to long to be on the wing. After this day they were missing at once; nor could I ever observe them with their dam coursing round the church in the act of learning to fly, as the first broods evidently do. On the thirty-first I caused the eaves to be searched, but we found in the nest only two callow, dead, stinking swifts, on which a second nest had been formed. This double nest was full of the black shining cases of the *hippoboscae hirundinis*.*

The following remarks on this unusual incident are obvious. The first is, that though it may be disagreeable to swifts to remain beyond the beginning of August, yet that they can subsist longer is undeniable. The second is, that this uncommon event, as it was owing to the loss of the first brood, so it corroborates my former remark, that swifts breed regularly but once; since, was the contrary the case, the occurrence above could neither be new nor rare.

P.S. One swift was seen at Lyndon, in the county of Rutland, in 1782, so late as the third of September.*

LETTER 53

As I have sometimes known you make inquiries about several kinds of insects, I shall here send you an account of one sort which I little expected to have found in this kingdom. I had often observed that one particular part of a vine growing on the walls of my house was covered in the autumn with a black dust-like appearance, on which the flies fed eagerly; and that the shoots and leaves thus affected did not thrive; nor did the fruit ripen. To this substance I applied my glasses; but could not discover that it had any thing to do with animal life, as I at first expected: but, upon a closer examination behind the larger boughs, we were surprised to find that they were coated over with husky shells, from whose sides proceeded a cotton-like substance, surrounding a multitude of eggs. This curious and uncommon production put me upon recollecting what I have heard and read concerning the *coccus vitis viniferae* of Linnaeus, which, in the south of Europe, infests many vines, and is an horrid and loathsome pest. As soon as I had turned to the accounts given of this insect, I saw at once that it swarmed on my vine; and did not appear to have been at all checked by the preceding winter, which had been uncommonly severe.

Not being then at all aware that it had any thing to do with England, I was much inclined to think that it came from Gibraltar among the many boxes and packages of plants and birds which I had formerly received from thence; and especially as the vine infested grew immediately under my study-window, where I usually kept my specimens. True it is that I had received nothing from thence for some years:* but as insects, we know, are conveyed from one country to another in a very unexpected manner, and have a wonderful power of maintaining their existence till they fall into a nidus* proper for their support and increase, I cannot but suspect still that these cocci came to me originally from Andalusia. Yet, all the while, candour obliges me to confess that Mr Lightfoot has written me word that he once, and but once, saw these insects on a vine at Weymouth in Dorsetshire;* which, it is here to be observed, is a sea-port town to which the coccus might be conveyed by shipping.

As many of my readers may possibly never have heard of this strange and unusual insect, I shall here transcribe a passage from a natural history of Gibraltar, written by the Reverend John White, late vicar of Blackburn in Lancashire, but not yet published:*—'In the year 1770 a vine, which grew on the east-side of my house, and which had produced the finest crops of grapes for years past, was suddenly overspread on all the woody branches with large lumps of a white fibrous substance resembling spiders webs, or rather raw cotton. It was of a very clammy quality, sticking fast to every thing that touched it, and capable of being spun into long threads. At first I suspected it to be the product of spiders, but could find none. Nothing was to be seen connected with it but many brown oval husky shells, which by no means looked like insects, but rather resembled bits of the dry bark of the vine. The tree had a plentiful crop of grapes set, when this pest appeared upon it; but the fruit was manifestly injured by this foul incumbrance. It remained all the summer, still increasing, and loaded the woody and bearing branches to a vast degree. I often pulled off great quantities by handfuls; but it was so slimy and tenacious that it could by no means be cleared. The grapes never filled to their natural perfection, but turned watery and vapid. Upon perusing the works afterwards of M. de Reaumur, I found this matter perfectly described and accounted for.* Those husky shells, which I had observed, were no other than the female coccus, from whose sides this cotton-like substance exsudes, and serves as a covering and security for their eggs.'

To this account I think proper to add, that, though the female cocci are stationary, and seldom remove from the place to which they stick, yet the male is a winged insect; and that the black dust which I saw was undoubtedly the excrement of the females, which is eaten by ants as well as flies. Though the utmost severity of our winter did not destroy these insects, yet the attention of the gardener* in a summer or two has entirely relieved my vine from this filthy annoyance.

As we have remarked above that insects are often conveyed from one country to another in a very unaccountable manner, I shall here mention an emigration of small aphides, which was observed in the village of Selborne no longer ago than August the 1st, 1785.

At about three o'clock in the afternoon of that day, which was very hot, the people of this village were surprised by a shower of aphides, or smother-flies, which fell in these parts. Those that were walking in the street at that juncture found themselves covered with these insects, which settled also on the hedges and gardens, blackening all the vegetables where they alighted. My annuals were discoloured with them, and the stalks of a bed of onions were quite coated over for six days after. These armies were then, no doubt, in a state of emigration, and shifting their quarters; and might have come, as far as we know, from the great hop-plantations of Kent or Sussex, the wind being all that day in the easterly quarter. They were observed at the same time in great clouds about Farnham, and all along the vale from Farnham to Alton.[1]

LETTER 54

Dear Sir,

When I happen to visit a family where gold and silver fishes are kept in a glass bowl,* I am always pleased with the occurrence, because it offers me an opportunity of observing the actions and propensities of those beings with whom we can be little acquainted in their natural state. Not long since I spent a fortnight at the house of a friend where there was such a vivary,* to which I paid no small attention, taking every occasion to remark what passed within its narrow limits. It was here that I first observed the manner in which fishes die. As soon as the creature sickens, the head sinks lower and lower, and it stands as it were on its head; till, getting weaker, and losing all poise, the tail turns over, and at last it floats on the surface of the water with its belly uppermost. The reason why fishes, when dead, swim in that manner is very obvious; because, when the body is no longer balanced by the fins of the belly, the broad muscular back preponderates by its own gravity, and turns the belly uppermost, as lighter from its being a cavity, and because it contains the swimming-bladders, which contribute to render it buoyant. Some that delight in gold and silver fishes have adopted a notion that they need no aliment.* True it is that they will subsist for a long time without any apparent food but what they can collect from pure water frequently changed; yet they must draw some support from animalcula,* and

---

[1] For various methods by which several insects shift their quarters, see Derham's *Physico-Theology*.

other nourishment supplied by the water; because, though they seem to eat nothing, yet the consequences of eating often drop from them. That they are best pleased with such jejune diet may easily be confuted, since if you toss them crumbs they will seize them with great readiness, not to say greediness: however, bread should be given sparingly, lest, turning sour, it corrupt* the water. They will also feed on the water-plant called *lemna* (duck's meat), and also on small fry.

When they want to move a little they gently protrude themselves with the *pinnae pectorales* [pectoral fins]; but it is with their strong muscular tails only that they and all fishes shoot along with such inconceivable rapidity. It has been said that the eyes of fishes are immoveable: but these apparently turn them forward or backward in their sockets as their occasions require. They take little notice of a lighted candle, though applied close to their heads, but flounce and seem much frightened by a sudden stroke of the hand against the support whereon the bowl is hung; especially when they have been motionless, and are perhaps asleep. As fishes have no eyelids, it is not easy to discern when they are sleeping or not, because their eyes are always open.

Nothing can be more amusing than a glass bowl containing such fishes: the double refractions of the glass and water represent them, when moving, in a shifting and changeable variety of dimensions, shades, and colours; while the two mediums, assisted by the concavo-convex shape of the vessel, magnify and distort them vastly; not to mention that the introduction of another element and its inhabitants into our parlours engages the fancy in a very agreeable manner.

Gold and silver fishes, though originally natives of China and Japan, yet are become so well reconciled to our climate as to thrive and multiply very fast in our ponds and stews.* Linnaeus ranks this species of fish under the genus of *cyprinus*, or carp, and calls it *cyprinus auratus*.

Some people exhibit this sort of fish in a very fanciful way; for they cause a glass bowl to be blown with a large hollow space within, that does not communicate with it. In this cavity they put a bird occasionally; so that you may see a goldfinch or a linnet hopping as it were in the midst of the water, and the fishes swimming in a circle round it. The simple exhibition of the fishes is agreeable and pleasant; but in so

complicated a way becomes whimsical and unnatural, and liable to
the objection due to him,

> Qui variare cupit rem prodigialitèr unam.
>
> [who delights to speculate on a subject that
> will scarcely bear it.]   [Horace, *Ars Poetica*]

I am, &c.

LETTER 55

October 10, 1781

Dear Sir,

I think I have observed before that much the most considerable
part of the house-martins withdraw from hence about the first week
in October; but that some, the latter broods I am now convinced,
linger on till towards the middle of that month: and that at times,
once perhaps in two or three years, a flight, for one day only, has
shown itself in the first week in November.

Having taken notice, in October 1780, that the last flight was
numerous, amounting perhaps to one hundred and fifty; and that the
season was soft and still; I was resolved to pay uncommon attention to
these late birds; to find, if possible, where they roosted, and to determine the precise time of their retreat. The mode of life of these latter
hirundines is very favourable to such a design; for they spend the
whole day in the sheltered district, between me and the Hanger, sailing about in a placid, easy manner, and feasting on those insects which
love to haunt a spot so secure from ruffling winds. As my principal
object was to discover the place of their roosting, I took care to wait on
them before they retired to rest, and was much pleased to find that, for
several evenings together, just at a quarter past five in the afternoon,
they all scudded away in great haste towards the south-east, and darted
down among the low shrubs above the cottages at the end of the hill.
This spot in many respects seems to be well calculated for their winter
residence: for in many parts it is as steep as the roof of any house, and
therefore secure from the annoyances of water; and it is moreover
clothed with beechen shrubs, which, being stunted and bitten by
sheep, make the thickest covert imaginable; and are so entangled as to

be impervious to the smallest spaniel: besides, it is the nature of under-wood beech never to cast its leaf all the winter; so that, with the leaves on the ground and those on the twigs, no shelter can be more complete. I watched them on to the thirteenth and fourteenth of October, and found their evening retreat was exact and uniform; but after this they made no regular appearance. Now and then a straggler was seen; and, on the twenty-second of October, I observed two in the morning over the village, and with them my remarks for the season ended.

From all these circumstances put together, it is more than probable that this lingering flight, at so late a season of the year, never departed from the island. Had they indulged me that autumn with a November visit, as I much desired, I presume that, with proper assistants, I should have settled the matter past all doubt; but though the third of November was a sweet day, and in appearance exactly suited to my wishes, yet not a martin was to be seen; and so I was forced, reluctantly, to give up the pursuit.

I have only to add that were the bushes, which cover some acres, and are not my own property, to be grubbed and carefully examined, probably those late broods, and perhaps the whole aggregate body of the house-martins of this district, might be found there, in different secret dormitories; and that, so far from withdrawing into warmer climes, it would appear that they never depart three hundred yards from the village.

LETTER 56

They who write on natural history cannot too frequently advert to instinct, that wonderful limited faculty, which, in some instances, raises the brute creation as it were above reason, and in others leaves them so far below it. Philosophers have defined instinct to be that secret influence by which every species is impelled naturally to pursue, at all times, the same way or track, without any teaching or example; whereas reason, without instruction, would often vary and do that by many methods which instinct effects by one alone. Now this maxim must be taken in a qualified sense; for there are instances in which instinct does vary and conform to the circumstances of place and convenience.

It has been remarked that every species of bird has a mode of nidification peculiar to itself; so that a school-boy would at once pronounce on the sort of nest before him. This is the case among fields
and woods, and wilds; but, in the villages round London, where
mosses and gossamer, and cotton from vegetables,* are hardly to be
found, the nest of the chaffinch has not that elegant finished appearance, nor is it so beautifully studded with lichens, as in a more rural
district: and the wren is obliged to construct its house with straws
and dry grasses, which do not give it that rotundity and compactness
so remarkable in the edifices of that little architect. Again, the regular
nest of the house-martin is hemispheric; but where a rafter, or a
joist, or a cornice, may happen to stand in the way, the nest is so con-
trived as to conform to the obstruction, and becomes flat or oval, or
compressed.

In the following instances instinct is perfectly uniform and consistent. There are three creatures, the squirrel, the field-mouse, and
the bird called the nut-hatch, (*sitta Europaea*), which live much on
hazelnuts; and yet they open them each in a different way. The first,
after rasping off the small end, splits the shell in two with his long
fore-teeth, as a man does with his knife; the second nibbles a hole
with his teeth, so regular as if drilled with a wimble,* and yet so small
that one would wonder how the kernel can be extracted through it;
while the last picks an irregular ragged hole with its bill: but as this
artist has no paws to hold the nut firm while he pierces it, like an
adroit workman, he fixes it, as it were in a vice, in some cleft of a tree,
or in some crevice; when, standing over it, he perforates the stubborn
shell. We have often placed nuts in the chink of a gate-post where
nut-hatches have been known to haunt, and have always found that
those birds have readily penetrated them. While at work they make
a rapping noise that may be heard at a considerable distance.

You that understand both the theory and practical part of music*
may best inform us why harmony or melody should so strangely
affect some men, as it were by recollection, for days after a concert is
over. What I mean the following passage will most readily explain:

Praehabebat porrò vocibus humanis, instrumentisque harmonicis musi-
cam illam avium: non quod aliâ quoque non delectaretur; sed quod ex
musicâ humanâ relinqueretur in animo continens quaedam, atten-
tionemque et somnum conturbans agitatio; dum ascensus, exscensus, ten-
ores, ac mutationes illae sonorum, et consonantiarum euntque, redeuntque

per phantasiam:—cum nihil tale relinqui possit ex modulationibus avium, quae, quod non sunt perinde a nobis imitabiles, non possunt perinde internam facultatem commovere.

Gassendus in *Vitâ Peireskii**

[He preferred the music of birds to that made either by the human voice or by instruments. It was not that these last failed to give him pleasure; on the contrary, the melodies and harmony would run so repeatedly in his imagination that the continual pleasure distracted his mind and disturbed his sleep. Such, however, was not the case with the music of birds—for the very reason that these melodies do not lend themselves to be repeated by men and, hence, do not stay in the mind to vex and confuse the inner faculties.]

This curious quotation strikes me much by so well representing my own case, and by describing what I have so often felt, but never could so well express.* When I hear fine music I am haunted with passages therefrom night and day; and especially at first waking, which, by their importunity, give me more uneasiness than pleasure:* elegant lessons still tease my imagination, and recur irresistibly to my recollection at seasons, and even when I am desirous of thinking of more serious matters.

I am, &c.

LETTER 57

A rare, and I think a new, little bird frequents my garden, which I have great reason to think is the pettichaps: it is common in some parts of the kingdom; and I have received formerly several dead specimens from Gibraltar. This bird much resembles the white-throat, but has a more white or rather silvery breast and belly; is restless and active, like the willow-wrens, and hops from bough to bough, examining every part for food; it also runs up the stems of the crown-imperials, and, putting its head into the bells of those flowers, sips the liquor which stands in the nectarium* of each petal. Sometimes it feeds on the ground like the hedge-sparrow, by hopping about on the grass-plots and mown walks.

One of my neighbours, an intelligent and observing man, informs me that, in the beginning of May, and about ten minutes before eight o'clock in the evening, he discovered a great cluster of house-swallows, thirty at least he supposes, perching on a willow that hung over the verge of James Knight's upper-pond.* His attention was first drawn by the twittering of these birds, which sat motionless in a row on the bough, with their heads all one way, and, by their weight, pressing down the twig so that it nearly touched the water. In this situation he watched them till he could see no longer. Repeated accounts of this sort, spring and fall, induce us greatly to suspect that house-swallows have some strong attachment to water, independent of the matter of food; and, though they may not retire into that element, yet they may conceal themselves in the banks of pools and rivers during the uncomfortable months of winter.*

One of the keepers of Wolmer-forest sent me a peregrine falcon, which he shot on the verge of that district as it was devouring a wood-pigeon. The *falco peregrinus*, or haggard falcon, is a noble species of hawk seldom seen in the southern counties. In winter 1767 one was killed in the neighbouring parish of Faringdon, and sent by me to Mr Pennant into North-Wales.[1] Since that time I have met with none till now. The specimen mentioned above was in fine preservation, and not injured by the shot: it measured forty-two inches from wing to wing, and twenty-one from beak to tail, and weighed two pounds and an half standing weight. This species is very robust, and wonderfully formed for rapine: its breast was plump and muscular; its thighs long, thick, and brawny; and its legs remarkably short and well set: the feet were armed with most formidable, sharp, long talons: the eyelids and cere of the bill were yellow; but the irides of the eyes dusky; the beak was thick and hooked, and of a dark colour, and had a jagged process near the end of the upper mandible on each side: its tail, or train, was short in proportion to the bulk of its body: yet the wings, when closed, did not extend to the end of the train. From its large and fair proportions it might be supposed to have been a female; but I was not permitted to cut open the specimen. For one of the birds of prey, which are usually lean, this was in high case: in its craw were many barley-corns, which probably came from the crop of the wood-pigeon, on which it was feeding when shot: for voracious birds do not eat grain; but, when devouring their quarry, with undistinguishing vehemence swallow bones and feathers, and all matters,

---

[1] See my tenth and eleventh letter to that gentleman.

indiscriminately. This falcon was probably driven from the mountains of North Wales or Scotland, where they are known to breed, by rigorous weather and deep snows that had lately fallen.

I am, &c.

LETTER 58

My near neighbour, a young gentleman in the service of the East-India Company,* has brought home a dog and a bitch of the Chinese breed from Canton; such as are fattened in that country for the purpose of being eaten: they are about the size of a moderate spaniel; of a pale yellow colour, with coarse bristling hairs on their backs; sharp upright ears, and peaked heads, which give them a very fox-like appearance. Their hind legs are unusually straight without any bend at the hock or ham, to such a degree as to give them an aukward gait when they trot. When they are in motion their tails are curved high over their backs like those of some hounds, and have a bare place each on the outside from the tip midway, that does not seem to be matter of accident, but somewhat singular. Their eyes are jet-black, small, and piercing; the insides of their lips and mouths of the same colour, and their tongues blue. The bitch has a dew-claw on each hind leg; the dog has none. When taken out into a field the bitch showed some disposition for hunting, and dwelt on the scent of a covey of partridges till she sprung them, giving her tongue all the time. The dogs in South America are dumb;* but these bark much in a short thick manner, like foxes; and have a surly, savage demeanour like their ancestors, which are not domesticated, but bred up in sties, where they are fed for the table with rice-meal and other farinaceous food. These dogs, having been taken on board as soon as weaned, could not learn much from their dam; yet they did not relish flesh when they came to England. In the islands of the pacific ocean the dogs are bred up on vegetables, and would not eat flesh when offered them by our circumnavigators.

We believe that all dogs, in a state of nature, have sharp, upright fox-like ears; and that hanging ears, which are esteemed so graceful, are the effect of choice breeding and cultivation. Thus, in the *Travels* of Ysbrandt Ides from Muscovy to China, the dogs which draw the Tartars on snow-sledges near the river Oby are engraved with prick-ears,* like those from Canton. The Kamschatdales also train the same sort of sharp-eared peaked-nosed dogs to draw their sledges; as may be seen in an elegant print engraved for Captain Cook's last voyage round the world.*

Now we are upon the subject of dogs, it may not be impertinent to add, that spaniels, as all sportsmen know, though they hunt partridges and pheasants as it were by instinct, and with much delight and alacrity, yet will hardly touch their bones when offered as food; nor will a mongrel dog of my own, though he is remarkable for finding that sort of game. But, when we came to offer the bones of partridges to the two Chinese dogs, they devoured them with much greediness, and licked the platter clean.

No sporting dogs will flush woodcocks till inured to the scent and trained to the sport, which they then pursue with vehemence and transport; but then they will not touch their bones, but turn from them with abhorrence, even when they are hungry.

Now, that dogs should not be fond of the bones of such birds as they are not disposed to hunt is no wonder; but why they reject and do not care to eat their natural game is not so easily accounted for, since the end of hunting seems to be, that the chase pursued should be eaten. Dogs again will not devour the more rancid water-fowls, nor indeed the bones of any wild-fowls; nor will they touch the foetid bodies of birds that feed on offal and garbage: and indeed there may be somewhat of providential instinct in this circumstance of dislike; for vultures,[1] and kites, and ravens, and crows, &c. were intended to be messmates with dogs[2] over their carrion; and seem to be appointed by Nature as fellow-scavengers to remove all cadaverous nuisances from the face of the earth.

I am, &c.

[1] Hasselquist, in his *Travels to the Levant*, observes that the dogs and vultures at Grand Cairo maintain such a friendly intercourse as to bring up their young together in the same place.

[2] The Chinese word for a dog to an European ear sounds like quihloh.

LETTER 59

The fossil wood buried in the bogs of Wolmer-forest is not yet all exhausted; for the peat-cutters now and then stumble upon a log. I have just seen a piece which was sent by a labourer of Oakhanger to a carpenter of this village; this was the but-end of a small oak, about five feet long, and about five inches in diameter. It had apparently been severed from the ground by an axe, was very ponderous, and as black as ebony. Upon asking the carpenter for what purpose he had procured it; he told me that it was to be sent to his brother, a joiner at Farnham, who was to make use of it in cabinet work, by inlaying it along with whiter woods.

Those that are much abroad on evenings after it is dark, in spring and summer, frequently hear a nocturnal bird passing by on the wing, and repeating often a short quick note. This bird I have remarked myself, but never could make out till lately. I am assured now that it is the Stone-curlew, (*charadrius oedicnemus*). Some of them pass over or near my house almost every evening after it is dark,* from the uplands of the hill and North field, away down towards Dorton; where, among the streams and meadows, they find a greater plenty of food. Birds that fly by night are obliged to be noisy; their notes often repeated become signals or watch-words to keep them together, that they may not stray or lose each the other in the dark.

The evening proceedings and manoeuvres of the rooks are curious and amusing in the autumn. Just before dusk they return in long strings from the foraging of the day, and rendezvous by thousands over Selborne-down, where they wheel round in the air, and sport and dive in a playful manner, all the while exerting their voices, and making a loud cawing, which, being blended and softened by the distance that we at the village are below them, becomes a confused noise or chiding; or rather a pleasing murmur, very engaging to the imagination, and not unlike the cry of a pack of hounds in hollow, echoing woods, or the rushing of the wind in tall trees, or the tumbling of the tide upon a pebbly shore. When this ceremony is over, with the last gleam of day, they retire for the night to the deep beechen woods of Tisted and Ropley. We remember a little girl who, as she was going to bed, used to remark on such an occurrence, in the true spirit of physico-theology, that the rooks were saying their prayers; and yet this child was much too young to be aware that the scriptures have said of the Deity—that 'he feedeth the ravens who call upon him.'*

I am, &c.

LETTER 60

In reading Dr Huxham's *Observationes de Aëre,** &c. written at Plymouth, I find by those curious and accurate remarks, which contain an account of the weather from the year 1727 to the year 1748, inclusive, that though there is frequent rain in that district of Devonshire, yet the quantity falling is not great; and that some years it has been very small: for in 1731 the rain measured only 17^{inch}.—266^{thou}. and in 1741, 20—354; and again, in 1743 only 20—908. Places near the sea have frequent scuds,* that keep the atmosphere moist, yet do not reach far up into the country; making thus the maritime situations appear wet, when the rain is not considerable. In the wettest years at Plymouth the Doctor measured only once 36; and again once, *viz.* 1734, 37—114: a quantity of rain that has twice been exceeded at Selborne in the short period of my observations. Dr Huxham remarks that frequent small rains keep the air moist; while heavy ones render it more dry, by beating down the vapours. He is also of opinion that the dingy, smoky appearance in the sky, in very dry seasons, arises from the want of moisture sufficient to let the light through, and render the atmosphere transparent; because he had observed several bodies more diaphanous when wet than dry; and did never recollect that the air had that look in rainy seasons.

My friend, who lives just beyond the top of the down, brought his three swivel guns* to try them in my outlet, with their muzzles towards the Hanger, supposing that the report would have had a great effect; but the experiment did not answer his expectation. He then removed them to the Alcove on the Hanger; when the sound, rushing along the Lythe and Comb-wood, was very grand: but it was at the Hermitage* that the echoes and repercussions delighted the hearers; not only filling the Lythe with the roar, as if all the beeches were tearing up by the roots; but, turning to the left, they pervaded the vale above Combwood-ponds; and after a pause seemed to take up the crash again, and to extend round Harteley-hangers, and to die away at last among the coppices and coverts of Ward-le-ham. It has been remarked before that this district is an anathoth,* a place of responses or echoes, and therefore proper for such experiments: we may farther add that the pauses in echoes, when they cease and yet are taken up again, like the pauses in music, surprise the hearers, and have a fine effect on the imagination.

The gentleman abovementioned has just fixed a barometer in his parlour at Newton Valence.* The tube was first filled here (at Selborne) twice with care, when the mercury agreed and stood exactly with my own; but, being filled again twice at Newton, the mercury stood, on account of the great elevation of that house, three-tenths of an inch lower than the barometers at this village, and so continues to do, be the weight of the atmosphere what it may. The plate of the barometer at Newton is figured as low as 27; because in stormy weather the mercury there will sometimes descend below 28. We have supposed Newton-house to stand two hundred feet higher than this house: but if the rule holds good, which says that mercury in a barometer sinks one-tenth of an inch for every hundred feet elevation, then the Newton barometer, by standing three-tenths lower than that of Selborne, proves that Newton-house must be three hundred feet higher than that in which I am writing, instead of two hundred.

It may not be impertinent to add, that the barometers at Selborne stand three-tenths of an inch lower than the barometers at South Lambeth: whence we may conclude that the former place is about three hundred feet higher than the latter; and with good reason because the streams that rise with us run into the Thames at Weybridge, and so to London. Of course therefore there must be lower ground all the way from Selborne to South Lambeth; the distance between which, all the windings and indentings of the streams considered, cannot be less than an hundred miles.

I am, &c.

LETTER 61

Since the weather of a district is undoubtedly part of its natural history, I shall make no further apology for the four following letters,* which will contain many particulars concerning some of the great frosts and a few respecting some very hot summers, that have distinguished themselves from the rest during the course of my observations.

As the frost in January 1768 was, for the small time it lasted, the most severe that we had then known for many years, and was remarkably injurious to ever-greens, some account of its rigour, and reason of its ravages, may be useful, and not unacceptable to persons that delight in planting and ornamenting; and may particularly become a work that professes never to lose sight of utility.

For the last two or three days of the former year there were considerable falls of snow, which lay deep and uniform on the ground without any drifting, wrapping up the more humble vegetation in perfect security. From the first day to the fifth of the new year more snow succeeded; but from that day the air became entirely clear; and the heat of the sun about noon had a considerable influence in sheltered situations.

It was in such an aspect that the snow on the author's evergreens was melted every day, and frozen intensely every night; so that the laurustines, bays, laurels, and arbutuses looked, in three or four days, as if they had been burnt in the fire; while a neighbour's plantation of the same kind, in a high cold situation, where the snow was never melted at all, remained uninjured.

From hence I would infer that it is the repeated melting and freezing of the snow that is so fatal to vegetation, rather than the severity of the cold. Therefore it highly behoves every planter, who wishes to escape the cruel mortification of losing in a few days the labour and hopes of years, to bestir himself on such emergencies; and, if his plantations are small, to avail himself of mats, cloths, pease-haum,* straw, reeds, or any such covering, for a short time; or, if his shrubberies are extensive, to see that his people go about with prongs and forks, and carefully dislodge the snow from the boughs: since the naked foliage will shift much better for itself, than where the snow is partly melted and frozen again.

It may perhaps appear at first like a paradox; but doubtless the more tender trees and shrubs should never be planted in hot aspects; not only for the reason assigned above, but also because, thus circumstanced, they are disposed to shoot earlier in the spring, and to grow on later in the autumn, than they would otherwise do, and so are sufferers by lagging or early frosts. For this reason also plants from Siberia will hardly endure our climate; because, on the very first advances of spring, they shoot away, and so are cut off by the severe nights of March or April.

Dr Fothergill and others have experienced the same inconvenience with respect to the more tender shrubs from North-America; which they therefore plant under north-walls. There should also perhaps be a wall to the east to defend them from the piercing blasts from that quarter.

This observation might without any impropriety be carried into animal life; for discerning bee-masters now find that their hives should not in the winter be exposed to the hot sun, because such unseasonable warmth awakens the inhabitants too early from their slumbers; and, by putting their juices into motion too soon, subjects them afterwards to inconveniencies when rigorous weather returns.

The coincidents attending this short but intense frost were, that the horses fell sick with an epidemic distemper, which injured the winds* of many, and killed some; that colds and coughs were general among the human species; that it froze under people's beds for several nights;* that meat was so hard frozen that it could not be spitted, and could not be secured but in cellars; that several redwings and thrushes were killed by the frost; and that the large titmouse continued to pull straws lengthwise from the eaves of thatched houses and barns in a most adroit manner, for a purpose that has been explained already.[1]

On the 3d of January Benjamin Martin's thermometer within doors, in a close parlour where there was no fire, fell in the night to 20, and on the 4th to 18, and on the 7th to 17½, a degree of cold which the owner never since saw in the same situation; and he regrets much that he was not able at that juncture to attend his instrument abroad. All this time the wind continued north and north-east; and yet on the 8th roost-cocks, which had been silent, began to sound their clarions, and crows to clamour, as prognostic of milder weather; and, moreover, moles began to heave and work, and a manifest thaw took place. From the latter circumstance we may conclude that thaws often originate under ground from warm vapours which arise; else how should subterraneous animals receive such early intimations of their approach. Moreover, we have often observed that cold seems to

[1] See Letter 41 to Mr Pennant.

descend from above; for, when a thermometer hangs abroad in a frosty night, the intervention of a cloud shall immediately raise the mercury ten degrees; and a clear sky shall again compel it to descend to its former gauge.

And here it may be proper to observe, on what has been said above, that though frosts advance to their utmost severity by somewhat of a regular gradation, yet thaws do not usually come on by as regular a declension* of cold; but often take place immediately from intense freezing; as men in sickness often mend at once from a paroxysm.

To the great credit of Portugal laurels and American junipers, be it remembered that they remained untouched amidst the general havock: hence men should learn to ornament chiefly with such trees as are able to withstand accidental severities, and not subject themselves to the vexation of a loss which may befall them once perhaps in ten years, yet may hardly be recovered through the whole course of their lives.

As it appeared afterwards the ilexes were much injured, the cypresses were half destroyed, the arbutuses lingered on, but never recovered; and the bays, laurustines, and laurels, were killed to the ground; and the very wild hollies, in hot aspects, were so much affected that they cast all their leaves.

By the 14th of January the snow was entirely gone; the turnips emerged not damaged at all, save in sunny places; the wheat looked delicately, and the garden plants were well preserved; for snow is the most kindly mantle that infant vegetation can be wrapped in: were it not for that friendly meteor no vegetable life could exist at all in northerly regions. Yet in Sweden the earth in April is not divested of snow for more than a fortnight before the face of the country is covered with flowers.

LETTER 62

There were some circumstances attending the remarkable frost in January 1776 so singular and striking, that a short detail of them may not be unacceptable.

The most certain way to be exact will be to copy the passages from my journal, which were taken from time to time as things occurred.

But it may be proper previously to remark that the first week in January was uncommonly wet, and drowned with vast rains from every quarter: from whence may be inferred, as there is great reason to believe is the case, that intense frosts seldom take place till the earth is perfectly glutted and chilled with water;[1] and hence dry autumns are seldom followed by rigorous winters.

January 7th.—Snow driving all the day, which was followed by frost, sleet, and some snow, till the 12th, when a prodigious mass overwhelmed all the works of men, drifting over the tops of the gates and filling the hollow lanes.

On the 14th the writer was obliged to be much abroad; and thinks he never before or since has encountered such rugged Siberian weather. Many of the narrow roads were now filled above the tops of the hedges; through which the snow was driven into most romantic and grotesque shapes, so striking to the imagination as not to be seen without wonder and pleasure. The poultry dared not to stir out of their roosting places; for cocks and hens are so dazzled and confounded by the glare of snow that they would soon perish without assistance. The hares also lay sullenly in their seats, and would not move till compelled by hunger; being conscious, poor animals, that the drifts and heaps treacherously betray their footsteps, and prove fatal to numbers of them.

From the 14th the snow continued to increase, and began to stop the road waggons and coaches, which could no longer keep on their regular stages; and especially on the western roads, where the fall appears to have been deeper than in the south. The company at Bath, that wanted to attend the Queen's birth-day, were strangely incommoded: many carriages of persons, who got in their way to town from Bath as far as Marlborough, after strange embarrassments, here met with a *ne plus ultra* [an impasse]. The ladies fretted, and offered large rewards to labourers if they would shovel them a track to London: but the relentless heaps of snow were too bulky to be removed; and so the 18th passed over,* leaving the company in very uncomfortable circumstances at the Castle and other inns.

On the 20th the sun shone out for the first time since the frost

---

[1] The autumn preceding January 1768 was very wet, and particularly the month of September, during which there fell at Lyndon, in the county of Rutland, six inches and an half of rain. And the terrible long frost in 1739–40 set in after a rainy season, and when the springs were very high.

began; a circumstance that has been remarked before much in favour of vegetation. All this time the cold was not very intense, for the thermometer stood at 29, 28, 25, and thereabout; but on the 21st it descended to 20. The birds now began to be in a very pitiable and starving condition. Tamed by the season, sky-larks settled in the streets of towns, because they saw the ground was bare; rooks frequented dunghills close to houses; and crows watched horses as they passed, and greedily devoured what dropped from them; hares now came into men's gardens, and, scraping away the snow, devoured such plants as they could find.

On the 22d the author had occasion to go to London through a sort of Laplandian-scene, very wild and grotesque indeed. But the metropolis itself exhibited a still more singular appearance than the country; for, being bedded deep in snow, the pavement of the streets could not be touched by the wheels or the horses' feet, so that the carriages ran about without the least noise. Such an exemption from din and clatter was strange, but not pleasant; it seemed to convey an uncomfortable idea of desolation:

. . . ipsa silentia terrent.
[the very silence terrifies.]
[Virgil, *Aeneid* 2]

On the 27th much snow fell all day, and in the evening the frost became very intense. At South Lambeth, for the four following nights, the thermometer fell to 11, 7, 6, 6; and at Selborne to 7, 6, 10; and on the 31st of January, just before sun-rise, with rime* on the trees and on the tube of the glass, the quicksilver sunk exactly to zero, being 32 degrees below the freezing point: but by eleven in the morning, though in the shade, it sprung up to 16½.[1]—a most unusual degree of cold this for the south of England! During these four nights the cold was so penetrating that it occasioned ice in warm chambers and under beds; and in the day the wind was so keen that persons of robust constitutions could scarcely endure to face it. The Thames was at once so frozen over both above and below bridge that crowds ran about on the ice. The streets were now strangely encumbered

---

[1] At Selborne the cold was greater than at any other place that the author could hear of with certainty: though some reported at the time that at a village in Kent the thermometer fell two degrees below zero, *viz.* 34 degrees below the freezing point.

The thermometer used at Selborne was graduated by Benjamin Martin.

with snow, which crumbled and trod dusty; and, turning grey, resembled bay-salt: what had fallen on the roofs was so perfectly dry that, from first to last, it lay twenty-six days on the houses in the city; a longer time than had been remembered by the oldest housekeepers living. According to all appearances we might now have expected the continuance of this rigorous weather for weeks to come, since every night increased in severity; but behold, without any apparent cause, on the 1st of February a thaw took place, and some rain followed before night; making good the observation above, that frosts often go off as it were at once, without any gradual declension of cold. On the 2d of February the thaw persisted; and on the 3d swarms of little insects were frisking and sporting in a court-yard at South Lambeth, as if they had felt no frost. Why the juices in the small bodies and smaller limbs of such minute beings are not frozen is a matter of curious inquiry.

Severe frosts seem to be partial, or to run in currents; for, at the same juncture, as the author was informed by accurate correspondents, at Lyndon, in the county of Rutland, the thermometer stood at 19; at Blackburn, in Lancashire, at 19; and at Manchester at 21, 20, and 18. Thus does some unknown circumstance strangely overbalance latitude, and render the cold sometimes much greater in the southern than the northern parts of this kingdom.

The consequences of this severity were, that in Hampshire, at the melting of the snow, the wheat looked well, and the turnips came forth little injured. The laurels and laurustines were somewhat damaged, but only in hot aspects. No evergreens were quite destroyed; and not half the damage sustained that befell in January 1768. Those laurels that were a little scorched on the south-sides were perfectly untouched on their north-sides. The care taken to shake the snow day by day from the branches seemed greatly to avail the author's evergreens. A neighbour's laurel-hedge, in a high situation, and facing to the north, was perfectly green and vigorous;* and the Portugal laurels remained unhurt.

As to the birds, the thrushes and blackbirds were mostly destroyed; and the partridges, by the weather and poachers, were so thinned that few remained to breed the following year.

LETTER 63

As the frost in December 1784 was very extraordinary, you, I trust, will not be displeased to hear the particulars; and especially when I promise to say no more about the severities of winter after I have finished this letter.

The first week in December was very wet, with the barometer very low. On the 7th, with the barometer at 28—five tenths, came on a vast snow, which continued all that day and the next, and most part of the following night; so that by the morning of the 9th the works of men were quite overwhelmed, the lanes filled so as to be impassable, and the ground covered twelve or fifteen inches without any drifting. In the evening of the 9th the air began to be so very sharp that we thought it would be curious to attend to the motions of a thermometer: we therefore hung out two; one made by Martin and one by Dollond, which soon began to shew us what we were to expect; for, by ten o'clock, they fell to 21, and at eleven to 4, when we went to bed. On the 10th, in the morning, the quicksilver of Dollond's glass was down to half a degree below zero; and that of Martin's, which was absurdly graduated only to four degrees above zero, sunk quite into the brass guard of the ball; so that when the weather became most interesting this was useless. On the 10th, at eleven at night, though the air was perfectly still, Dollond's glass went down to one degree below zero! This strange severity of the weather made me very desirous to know what degree of cold there might be in such an exalted and near situation as Newton. We had therefore, on the morning of the 10th, written to Mr ——, and entreated him to hang out his thermometer, made by Adams; and to pay some attention to it morning and evening; expecting wonderful phaenomena, in so elevated a region, at two hundred feet or more above my house. But, behold! on the 10th, at eleven at night, it was down only to 17, and the next morning at 22, when mine was at ten! We were so disturbed at this unexpected reverse of comparative local cold, that we sent one of my glasses up, thinking that of Mr —— must, some how, be wrongly constructed. But, when the instruments came to be confronted, they went exactly together: so that, for one night at least, the cold at Newton was 18 degrees less than at Selborne; and, through the whole frost, 10 or 12 degrees; and indeed, when we came to observe consequences, we could readily credit this; for all my laurustines, bays, ilexes, arbutuses, cypresses, and even my Portugal laurels,[1] and (which occasions more regret) my fine sloping laurel-hedge, were scorched up; while, at Newton, the same trees have not lost a leaf!

We had steady frost on to the 25th, when the thermometer in the morning was down to 10 with us, and at Newton only to 21. Strong frost continued till the 31st, when some tendency to thaw was observed; and, by January the 3d, 1785, the thaw was confirmed, and some rain fell.

A circumstance that I must not omit, because it was new to us, is, that on Friday, December the 10th, being bright sun-shine, the air was full of icy spiculae,* floating in all directions, like atoms in a sun-beam let into a dark room. We thought them at first particles of the rime falling from my tall hedges; but were soon convinced to the contrary, by making our observations in open places where no rime could reach us. Were they watery particles of the air frozen as they floated; or were they evaporations from the snow frozen as they mounted?

We were much obliged to the thermometers for the early information they gave us; and hurried our apples, pears, onions, potatoes, &c. into the cellar, and warm closets; while those who had not, or neglected such warnings, lost all their store of roots and fruits, and had their very bread and cheese frozen.

I must not omit to tell you that, during those two Siberian days, my parlour-cat was so electric, that had a person stroked her, and been properly insulated, the shock might have been given to a whole circle of people.*

I forgot to mention before, that, during the two severe days, two men, who were tracing hares in the snow, had their feet frozen; and two men, who were much better employed, had their fingers so affected by the frost, while they were thrashing in a barn, that a mortification followed, from which they did not recover for many weeks.

This frost killed all the furze and most of the ivy, and in many places stripped the hollies of all their leaves. It came at a very early

---

[1] Mr Miller, in his *Gardener's Dictionary*, says positively that the Portugal laurels remained untouched in the remarkable frost of 1739–40. So that either that accurate observer was much mistaken, or else the frost of December 1784 was much more severe and destructive than that in the year above-mentioned.

time of the year, before old November* ended; and yet may be
allowed from its effects to have exceeded any since 1739–40.

LETTER 64

As the effects of heat are seldom very remarkable in the northerly
climate of England, where the summers are often so defective in
warmth and sun-shine as not to ripen the fruits of the earth so well as
might be wished, I shall be more concise in my account of the severity
of a summer season, and so make a little amends for the prolix account
of the degrees of cold, and the inconveniences that we suffered from
some late rigorous winters.

The summers of 1781 and 1783 were unusually hot and dry; to
them therefore I shall turn back in my journals, without recurring to
any more distant period. In the former of these years my peach and
nectarine-trees suffered so much from the heat that the rind on the
bodies was scalded and came off; since which the trees have been in a
decaying state. This may prove a hint to assiduous gardeners to fence
and shelter their wall-trees with mats or boards, as they may easily
do, because such annoyance is seldom of long continuance. During
that summer also, I observed that my apples were coddled,* as
it were, on the trees; so that they had no quickness of flavour, and
would not keep in the winter. This circumstance put me in mind of
what I have heard travellers assert, that they never ate a good apple or
apricot in the south of Europe, where the heats were so great as to
render the juices vapid and insipid.

The great pests of a garden are wasps, which destroy all the finer
fruits just as they are coming into perfection. In 1781 we had none; in
1783 there were myriads; which would have devoured all the produce
of my garden, had not we set the boys to take the nests, and caught
thousands with hazel-twigs tipped with bird-lime: we have since
employed the boys to take and destroy the large breeding wasps in the
spring. Such expedients have a great effect on these marauders, and
will keep them under. Though wasps do not abound but in hot summers, yet they do not prevail in every hot summer, as I have instanced
in the two years above-mentioned.

In the sultry season of 1783 honey-dews* were so frequent as to deface and destroy the beauties of my garden. My honeysuckles, which were one week the most sweet and lovely objects that the eye could behold, became the next the most loathsome; being enveloped in a viscous substance, and loaded with black aphides, or smother-flies. The occasion of this clammy appearance seems to be this, that in hot weather the effluvia of flowers in fields and meadows and gardens are drawn up in the day by a brisk evaporation, and then in the night fall down again with the dews, in which they are entangled; that the air is strongly scented, and therefore impregnated with the particles of flowers in summer weather, our senses will inform us; and that this clammy sweet substance is of the vegetable kind we may learn from bees, to whom it is very grateful: and we may be assured that it falls in the night, because it is always first seen in warm still mornings.

On chalky and sandy soils, and in the hot villages about London, the thermometer has been often observed to mount as high as 83 or 84; but with us, in this hilly and woody district, I have hardly ever seen it exceed 80; nor does it often arrive at that pitch. The reason, I conclude, is, that our dense clayey soil, so much shaded by trees, is not so easily heated through as those above-mentioned: and, besides, our mountains cause currents of air and breezes; and the vast effluvia from our woodlands temper and moderate our heats.

LETTER 65

The summer of the year 1783 was an amazing and portentous one, and full of horrible phaenomena; for, besides the alarming meteors* and tremendous thunder-storms that affrighted and distressed the different counties of this kingdom, the peculiar haze, or smokey fog, that prevailed for many weeks in this island, and in every part of Europe, and even beyond its limits, was a most extraordinary appearance, unlike any thing known within the memory of man.* By my journal I find that I had noticed this strange occurrence from June 23 to July 20 inclusive, during which period the wind varied to every quarter without making any alteration in the air. The sun, at noon, looked as blank as a clouded moon, and shed a rust-coloured ferruginous light on the ground, and floors of rooms; but was particularly lurid and blood-coloured at rising and setting. All the time the heat was so intense that butchers' meat could hardly be eaten on the day after it was killed; and the flies swarmed so in the lanes and hedges that they rendered the horses half frantic, and riding irksome. The country people began to look with a superstitious awe at the red, louring aspect of the sun; and indeed there was reason for the most enlightened person to be apprehensive; for, all the while, Calabria and part of the isle of Sicily, were torn and convulsed with earthquakes; and about that juncture a volcano sprung out of the sea on the coast of Norway. On this occasion Milton's noble simile of the sun, in his first book of *Paradise Lost*, frequently occurred to my mind; and it is indeed particularly applicable, because, towards the end, it alludes to a superstitious kind of dread, with which the minds of men are always impressed by such strange and unusual phaenomena.

> . . . As when the sun, new risen,
> Looks through the horizontal, misty air,
> Shorn of his beams; or from behind the moon,
> In dim eclipse, disastrous twilight sheds
> On half the nations, and with fear of change
> Perplexes monarchs . . .
>
> [Milton, *Paradise Lost*, I. 594–9]

LETTER 66

We are very seldom annoyed with thunder-storms: and it is no less remarkable than true, that those which arise in the south have hardly been known to reach this village; for, before they get over us, they take a direction to the east or to the west, or sometimes divide into two, and go in part to one of those quarters, and in part to the other; as was truly the case in summer 1783, when, though the country round was continually harassed with tempests, and often from the south, yet we escaped them all; as appears by my journal of that summer. The only way that I can at all account for this fact—for such it is—is that, on that quarter, between us and the sea, there are continual mountains, hill behind hill, such as Nore-hill, the Barnet, Butser-hill, and Portsdown, which some how divert the storms, and give them a different direction. High promontories, and elevated grounds, have always been observed to attract clouds and disarm them of their mischievous contents, which are discharged into the trees and summits as soon as they come in contact with those turbulent meteors; while the humble vales escape, because they are so far beneath them.

But, when I say I do not remember a thunder-storm from the south, I do not mean that we never have suffered from thunder-storms at all; for on June 5th, 1784, the thermometer in the morning being at 64, and at noon at 70, the barometer at 29—six tenths one-half, and the wind north, I observed a blue mist, smelling strongly of sulphur, hanging along our sloping woods, and seeming to indicate that thunder was at hand. I was called in about two in the afternoon, and so missed seeing the gathering of the clouds in the north; which they who were abroad assured me had something uncommon in its appearance. At about a quarter after two the storm began in the parish of Hartley, moving slowly from north to south; and from thence it came over Norton-farm, and so to Grange-farm, both in this parish. It began with vast drops of rain, which were soon succeeded by round hail, and then by convex pieces of ice, which measured three inches in girth. Had it been as extensive as it was violent, and of any con-tinuance (for it was very short), it must have ravaged all the neigh-bourhood. In the parish of Hartley it did some damage to one farm; but Norton, which lay in the center of the storm, was greatly injured; as was Grange, which lay next to it. It did but just reach to the middle of the village, where the hail broke my north windows, and all my garden-lights and hand-glasses,* and many of my neighbours' windows. The extent of the storm was about two miles in length and one in breadth. We were just sitting down to dinner; but were soon diverted from our repast by the clattering of tiles and the jingling of glass. There fell at the same time prodigious torrents of rain on the farms above-mentioned, which occasioned a flood as violent as it was sudden; doing great damage to the meadows and fallows, by deluging the one and washing away the soil of the other. The hollow lane towards Alton was so torn and disordered as not to be passable till mended, rocks being removed that weighed 200 weight. Those that saw the effect which the great hail had on ponds and pools say that the dashing of the water made an extraordinary appearance, the froth and spray standing up in the air three feet above the surface. The rushing and roaring of the hail, as it approached, was truly tremendous.

Though the clouds at South Lambeth, near London, were at that juncture thin and light, and no storm was in sight, nor within hearing, yet the air was strongly electric; for the bells of an electric machine at that place rang repeatedly, and fierce sparks were discharged.

When I first took the present work in hand I proposed to have added an 'Annus Historico-naturalis', or 'The Natural History of the Twelve Months of the Year'; which would have comprised many incidents and occurrences that have not fallen in my way to be mentioned in my series of letters;—but, as Mr Aikin of Warrington has lately published somewhat of this sort,* and as the length of my correspondence has sufficiently put your patience to the test, I shall here take a respectful leave of you and natural history together; And am,

> With all due deference and regard,
> Your most obliged,
> And most humble servant,
> GIL. WHITE

Selborne, June 25, 1787

APPENDIX I

STATISTICS OF SELBORNE

A STATE *of the* Parish *of* SELBORNE, *taken* October 4, 1783.

The number of tenements or families, 136.

The number of inhabitants in the street is 313
In the rest of the parish . . . 363 } Total, 676; near five inhabitants to each tenement.

In the time of the Rev. Gilbert White, vicar, who died in 1727–8, the number of inhabitants was computed at about 500.

Average of baptisms for 60 years.

From 1720 to 1729, both years inclusive	Males	6.9	12.9
	Females	6.0	
From 1730 to 1739, both years inclusive	Males	8.2	15.3
	Females	7.1	
From 1740 to 1749, inclusive	Males	9.2	15.8
	Females	6.6	
From 1750 to 1759, inclusive	Males	7.6	15.7
	Females	8.1	
From 1760 to 1769, inclusive	Males	9.1	18.0
	Females	8.9	
From 1770 to 1779, inclusive	Males	10.5	20.3
	Females	9.8	

Total baptisms of Males 515
Females 465 } 980

Total of baptisms from 1720 to 1779, both inclusive . . . 60 years . . . 980.

Average of burials for 60 years.

From 1720 to 1729, both years inclusive	Males	4.8	9.9
	Females	5.1	
From 1730 to 1739, both years inclusive	Males	4.8	10.6
	Females	5.8	
From 1740 to 1749, inclusive	Males	4.6	8.4
	Females	3.8	
From 1750 to 1759, inclusive	Males	4.9	10.0
	Females	5.1	

From 1760 to 1769, inclusive $\left\{\begin{array}{ll}\text{Males} & 6.9 \\ \text{Females} & 6.5\end{array}\right\}$ 13.4

From 1770 to 1779, inclusive $\left\{\begin{array}{ll}\text{Males} & 5.5 \\ \text{Females} & 6.2\end{array}\right\}$ 11.7

$\qquad$ Total of burials of Males $\quad$ 315 $\left.\right\}$ 640
$\qquad\qquad\qquad\qquad$ Females $\quad$ 325

Total of burials from 1720 to 1779 both inclusive ... 60 years ... 640.

Baptisms exceed burials by more than one third.
Baptisms of Males exceed Females by one tenth, or one in ten.
Burials of Females exceed Males by one in thirty.
It appears that a child, born and bred in this parish, has an equal chance
to live above forty years.
Twins thirteen times, many of whom dying young have lessened the
chance for life.
Chances for life in men and women appear to be equal.

A TABLE *of the* Baptisms, Burials, *and* Marriages, *from* January 2, 1761,
to December 25, 1780, *in the* Parish *of* SELBORNE.

	BAPTISMS			BURIALS			MAR.
	Males	Females	Total	Males	Females	Total	
1761	8	10	18	2	4	6	3
1762	7	8	15	10	14	24	6
1763	8	10	18	3	4	7	5
1764	11	9	20	10	8	18	6
1765	12	6	18	9	7	16	6
1766	9	13	22	10	6	16	4
1767	14	5	19	6	5	11	2
1768	7	6	13	2	5	7	6
1769	9	14	23	6	5	11	2
1770	10	13	23	4	7	11	3
1771	10	6	16	3	4	7	4
1772	11	10	21	6	10	16	3
1773	8	5	13	7	5	12	3
1774	6	13	19	2	8	10	1
1775	20	7	27	13	8	21	6
1776	11	10	21	4	6	10	6
1777	8	13	21	7	3	10	4
1778	7	13	20	3	4	7	5
1779	14	8	22	5	6	11	5
1780	8	9	17	11	4	15	3
	198	188	386	123	123	246	83

During this period of twenty years the births of males exceeded those of females . . . 10.

The burials of each sex were equal.

And the births exceeded the deaths . . . 140.

APPENDIX II

RESPONSES TO *THE NATURAL HISTORY OF SELBORNE*

I rejoyce excessively at your being now committed to the Press. As to your Frights & Fears, they become You well enough as a modest Man, but they are unnecessary as an Author. But I will put You in great Heart: Dr Chelsum told me that he had *seen your Book*, that it seem'd a very promising Performance, & likely to get into great Favour, that it was well put forth & decorated with very pleasing Prints & Views. Dr Chelsum is a Man of knowledge, a Connoscenti, & deep in Virtu. What would you have more in your prae-existing State? this is but your Embrio Glory; your material & Substantial happiness & Enjoyment is to come.

> Clergyman John Mulso, letter to Gilbert White, 21 July 1788, *The Letters to Gilbert White of Selborne from his Intimate Friend and Contemporary the Rev. John Mulso*, ed. Rashleigh Holt-White (London, 1907; facs. edn., Cambridge, 2011), 341.

Most of the local Histories which have fallen into our hands have been taken up with descriptions of the vestiges of ancient art and industry, while natural observations have been too much neglected. But we agree with Mr. White in his idea of parochial history, which, he thinks, ought to consist of natural productions and occurrences, as well as antiquities. For antiquities, when once surveyed, seldom recall farther attention, and are confined to one spot; whereas the pleasures of the naturalist continue through the year, return with unabated attractions every spring, and may be extended over the kingdom. . . .

If this author should be thought by any to have been too minute in his researches, let it be remembered that his studies have been in the great book of Nature. It must be confessed, that the oeconomy of the several kinds of crickets, and the distinction between the stock-dove and the ring-dove, are humble pursuits, and will be esteemed trivial by many, perhaps by some to be objects of ridicule. However, before we condemn any pursuits which contribute so much to health by calling us abroad, let us consider how the studious have employed themselves in their closets. In a former century the minds of the learned were engaged in determining whether the name of the Roman poet should be spelt Vergilius or Virgilius; and the number of letters in the name of Shakespeare still remains a matter of much solicitude and criticism. . . .

Disgusted as we have frequently been with writers who have told us only what they had read, with which we were already acquainted, and of which few were ignorant, it was with pleasure we perused this gentleman's account of the Natural History of his district, as he chiefly relates what he himself saw. The custom which has too long and too generally prevailed of compiling books from books has proved a great hindrance to the advancement of science.

> Thomas White, Gilbert White's brother, in his anonymous review of *The Natural History and Antiquities of Selborne* in *Gentleman's Magazine*, 59 (1789), 60–3, 144–6, at 60–1, 62, 144.

It is a part of our plan to notice all new publications, that illustrate the topography of the kingdom. And the first book upon this subject that it is our province to review, becomes a very pleasing task to us; for a more delightful, or more original work than Mr. White's History of Selborne has seldom been published. . . . Natural History has evidently been the author's principal study, and of that, ornithology is as evidently his favourite. The book is not a compilation from former publications; but the result of many years' attentive observations to nature itself, which are told not only with the precision of a philosopher, but with that happy selection of circumstances, which mark the *poet*. Throughout therefore, not only the understanding is informed, but the imagination touched.

> Review of *The Natural History and Antiquities of Selborne* in *The Topographer* 1 (1789), 40–4, at 40.

I have been greatly entertained by your 'Natural History of Selborn'. . . .

I agree with you, that naturalists in general attempt to explore too wide a field, their researches are too extensive; whereas if persons well qualified were to confine themselves to particular districts the natural history compiled from provincial authors would no doubt throw much light on the subject. . . .

Your work produced in me fresh ardor, and with that degree of enthusiasm necessary to such investigation, I pervaded the interior recesses of the thickest woods, and spread my researches to every place within my reach . . .

> Army officer and ornithologist, George Montagu, letters to White, 21 May and 27 June 1789, in Rashleigh Holt-White, *The Life and Letters of Gilbert White of Selborne*, 2 vols. (London, 1901), ii, 198, 201.

I remember reading White's *Natural History and Antiquities of Selborn*[e] with great pleasure when a Boy at school . . .

Poet William Wordsworth, letter to Francis Wrangham, 2 October 1808, *The Letters of William and Dorothy Wordsworth: The Middle Years, Part I: 1806–1811*, ed. Ernest De Selincourt, rev. Mary Moorman, 2nd edn. (Oxford, 1969), 270.

This is an inadequate explanation. I would rather say, that Instinct is the wisdom of the species, not of the Individual; but that let any circumstance occur regularly thro' many generations, that then its *every-time-felt* inconvenience would by little & little act thro' the blind sensations on the organic frame of the animals, till at length they were *born* wise in that respect. And by the same process do they lose their not *in*ate but *con*nate wisdom: thus Hens hatched in an artificial oven, as in Egypt, in 3 or 4 generations (the same process having been repeated in each) lose the instinct of Brooding. I trust, that this Note will not be considered as lessening the value of this sweet delightful Book.

Poet Samuel Taylor Coleridge, annotation dated 7 July 1810 in his copy of the 1802 edition of *The Natural History of Selborne* (letter 18 to Daines Barrington), *The Collected Works of Samuel Taylor Coleridge: Marginalia*, ed. H. J. Jackson and G. Whalley, 6 vols. (Princeton, 1980–2001), vi, 149.

From reading White's Selbourne I took much pleasure in watching the habits of birds & even made notes on this subject. In my simplicity I remember wondering why every gentleman did not become an ornithologist.

Naturalist Charles Darwin, recalling his school days, *c.*1819, in 'Recollections of my Mind and Character', *Charles Darwin: Evolutionary Writings*, ed. J. A. Secord, Oxford World's Classics (Oxford, 2008), p. 369.

I am reading, for the third time, 'White's Natural History of Selborne.' It is a book that would delight you, and be highly instructive to you in your art, if you are not already acquainted with it. White was the clergyman of the place, and occupied himself with narrowly observing and noting down all the natural occurrences that came within his view, and this for a number of years. It is most elegantly written. I fear the book is scarce. But if you can procure it, buy it for me and keep it by you. It is in your own way of close observation, and has in it that quality that, to me, constitutes the great pleasure of your society.

Archdeacon John Fisher, letter to the landscape artist John Constable, 6 March 1821, in C. R. Leslie, *The Life and Letters of John Constable, R. A.* (London, 1896), 97.

The mind that produced the 'Selborne' is such a one as I have always envied. The single page of the life of Mr. White leaves a more lasting impression on my mind than all that has been written of Charles V., or any other renowned hero. It shows what a real love of nature will do. Surely the serene and blameless life of Mr. White, so exempt from the folly and quackery of the world, must have fitted him for the clear and intimate view he took of nature. It proves the truth of Sir Joshua Reynolds's remark, 'that the virtuous man alone has true taste.' This book is an addition to my estate.

> Landscape artist John Constable, letter to Archdeacon John Fisher, 1 April 1821, in Leslie, *Life and Letters of John Constable*, 98–9.

I forgot to mention, that, in going from Hawkley to Greatham, the man, who went to show me the way, told me at a certain fork, 'that road goes to *Selborne.*' This put me in mind of a book, which was once recommended to me, but which I never saw, entitled '*The History and Antiquities of Selborne,*' (or something of that sort) written, I think, by a parson of the name of *White*, brother of Mr. *White*, so long a Bookseller in Fleet-street. This parson had, I think, the living of the parish of Selborne. The book was mentioned to me as a work of great curiosity and interest. But, at that time, the THING [the system of political corruption and economic injustice] was biting *so very sharply* that one had no attention to bestow on antiquarian researches. Wheat at 39*s.* a quarter, and Southdown ewes at 12*s.* 6*d.* have so weakened the THING'S jaws and so filed down its teeth, that I shall now certainly read this book if I can get it. By-the-bye if *all the parsons* had, for the last thirty years, employed their leisure time in writing the histories of their several parishes, instead of living, as many of them have, engaged in pursuits that I need not here name, neither their situation nor that of their flocks would, perhaps, have been the worse for it at this day.

> Radical journalist William Cobbett, writing in November 1822, in *Rural Rides . . . with Economical and Political Observations* (London, 1830), *97–*98.

Of all situations for a constant residence, that which appears to me most delightful is a little village far in the country; a small neighbourhood, not of fine mansions finely peopled, but of cottages and cottage-like houses . . . with inhabitants whose faces are as familiar to us as the flowers in our garden; a little world of our own, close-packed and insulated like ants in an ant-hill, or bees in a hive, or sheep in a fold, or nuns in a convent, or sailors in a ship; where we know every one, are known to every one, interested in every one, and authorised to hope that every one feels an interest in us.

How pleasant it is to slide into these fine-hearted feelings from the kindly and unconscious influence of habit, and to learn to know and to love the people about us, with all their peculiarities, just as we learn to know and to love the nooks and turns of the shady lanes and sunny commons that we pass every day. Even in books I like a confined locality . . . Nothing is so tiresome as to be whirled half over Europe at the chariot-wheels of a hero, to go to sleep at Vienna, and awaken at Madrid; it produces a real fatigue, a weariness of spirit. On the other hand, nothing is so delightful as to sit down in a country village in one of Miss Austen's delicious novels, quite sure before we leave it to become intimate with every spot and every person it contains; or to ramble with Mr. White[1] over his own parish of Selborne, and form a friendship with the fields and coppices, as well as with the birds, mice, and squirrels, who inhabit them. . . . And a small neighbourhood is as good in sober waking reality as in poetry or prose . . .

[1] White's 'Natural History and Antiquities of Selborne'; one of the most fascinating books ever written. I wonder that no naturalist has adopted the same plan.

> Writer and playwright Mary Russell Mitford, in *Our Village: Sketches of Rural Character and Scenery* (London, 1824), 1–2.

The Natural History of Selborne, has probably been supposed by many to be formed upon a more local and confined plan than it really is. In fact, the greater part of the observations are applicable to all that portion of the island in which he resided, and were, indeed, made in various places. Almost the only matter absolutely local is the account of the antiquities of the village of Selborne; and this seemed to stand so much apart, that, however well calculated to gratify the lovers of topographical studies, it was thought that its entire omission would be no loss to the work, considered as a publication in Natural History.

> John White (1765–1855), writing about his 'late respected relation' Gilbert White, in *The Natural History of Selborne*, 2 vols. (London, 1825), i, pp. v–vi [published in two octavo volumes].

White's delightful work is no longer shut up in a quarto. It is most pleasing to witness the exertions made by eminent writers of our time to produce food for the juvenile mind. Shall we be pardoned for observing, that the 'Natural History of Selborne' ought to have a place among the household books of every English family?

> Editor and biographer John Gibson Lockhart, writing anonymously in 'Life and Writings of Lucian', *Quarterly Review* (January 1828), 32–50, at 41.

Many years have now passed away since we were presented with that very interesting and amusing book, the 'Natural History of Selborne:' nor do I recollect any publication at all resembling it having since appeared. It early impressed on my mind an ardent love for all the ways and economy of nature, and I was thereby led to the constant observance of the rural objects around me. Accordingly, reflections have arisen, and notes been made, such as the reader will find them. . . . The meditations of separate naturalists in fields, in wilds, in woods, may yield a similarity of ideas; yet the different aspects under which the same things are viewed, and characters considered, afford infinite variety of description and narrative: mine, I confess, are but brief and slight sketches; plain observations of nature, the produce often of intervals of leisure and shattered health, affording no history of the country; a mere outline of rural things; the journal of a traveller through the inexhaustible regions of Nature.

> Botanist John Leonard Knapp, in *The Journal of a Naturalist* (London, 1829), pp. v–vi.

The attention that of late years has been devoted to the study of Natural History, and its importance to our commerce, manufactures, and domestic economy, must render every attempt to increase or simplify our knowledge of it at once praiseworthy and desirable; and, it is hoped, will be a sufficient apology for the reprint of a work which has already gone through several editions. . . . The work consists of a series of letters . . . written in a clear and elegant, yet somewhat popular style; containing very varied information upon most subjects connected with the Natural History of the age, being rather that of an extensive district than of a particular spot or village. . . .

This Edition is confined to the natural history of Selborne alone, including extracts from the author's Miscellaneous Observations, which are occasionally introduced as notes, to which the present Editor has subjoined such additional memoranda, as modern discoveries and the advanced state of knowledge rendered necessary.

As a naturalist, Mr White ranked very high . . . At the present time, we find his works extensively quoted, and many of his observations borne out by more modern experiments.

> Naturalist William Jardine, in the Introduction to his edition of *The Natural History of Selborne*, Constable's Miscellany, xlv (Edinburgh, 1829), extracts from pp. v–viii. This was the first inexpensive, popular edition of White's work.

The writings, and even the name of Gilbert White . . . have remained until lately almost unknown, save to the lovers of natural history, and to the curious few who, albeit themselves unskilled in such researches, had yet chanced to make a fond acquaintance with his unobtrusive pages. It is only in silence that a more general taste for their merits has been growing up; and the public are indebted to the conductors of 'Constable's Miscellany' for almost a new pleasure by the republication—though in too mutilated a shape—of the greater portion of the work. . . .

Among naturalists, Gilbert White, despite of the advancement of modern discoveries and science, still retains, we are told, a high station; but to us the chief attraction of his book lies in the moral beauty of his mind, in its serene and benevolent temper, in its deep, fervent feeling for all nature, in its reverential communing with the mysteries of creation, and, above all, in the perfect envelopement of its intellectual pleasures in that religion of the heart which sends up its incense to the Maker, even in the diurnal and hourly contemplation of his works. . . . To the admirers of Gilbert White, the scene which has been illustrated by his life and labours must ever be a fairy spot of the imagination . . . .

It has been excluded from the blessings of increasing commerce and population, from factories and filiations, manufacturers and Methodism, genius and gin, prosperity and pauperism. . . . Selborne has been condemned to remain unchanged amidst the changeful scenes of this stirring age . . .

'Selborne', *New Monthly Magazine*, 29 (1830), 564–70, at 564, 565, 566.

This work is written in an unconnected form, without any attempt at scientific arrangement, with which, however, Mr White shews himself well acquainted; and the minute exactness of his facts—the good taste displayed in their selection—and the elegance and liveliness with which they are described,—render this one of the most amusing books of the kind ever published, and it has gained for the author a high and just reputation.

Naturalist and army officer Captain Thomas Brown, in the Preface to his edition of *The Natural History of Selborne* (Edinburgh, 1833), pp. v–vi.

It is [White's] accuracy of observation, combined with a cheerful, benevolent, and pious spirit, which has at length rendered the Natural History of Selborne a book for all. . . . It is difficult to select a detached passage from the Natural History of Selborne that may give a proper idea of the merits of this delightful book. Nor is it necessary that we should do so; for the work itself ought to form a part of every library, and is one which we would especially recommend to all those who unite for the purchase of standard books.

'White's Natural History of Selborne', *Penny Magazine for the Society for the Diffusion of Useful Knowledge* (January 1833), 36–8, at 37.

I write with a request from my Sister [Dorothy Wordsworth], that you would bring down for her a Copy of Whites Natural History of Selborne which she has long wished to possess—The Book, as originally published was in Quarto entitled Natural History and Antiquities of Selborne, but the antiquities being of less general interest have not in many Editions been Reprinted. I think I have seen an Edition of the Natural History, alone, advertized, by Sir Wm Jardine, with Notes and a few Engravings—this is probably the Book which would best suit my Sister, and I believe it is not expensive.

Poet William Wordsworth, letter to Henry Crabb Robinson [? 24 May 1833], in *The Letters of Willam and Dorothy Wordsworth: The Later Years, Part II: 1829–1834*, ed. Ernest De Selincourt, 2nd edn., rev. A. G. Hill (Oxford, 1979), 622–3.

The beauty of White's Letters from Selborne does not consist in the curiosity or the remarkableness of his observations. There are many other works far richer in the wonders of the animal creation; and though the author never would have turned away from a *lusus naturæ*, his interest evidently lay in searching into her ordinary operations, in discovering those secrets which though hid from the vulgar eye, are easily unravelled by an observer who puts his heart in to the task of finding them out. . . . Stillness and quiet are the most decided characteristics of the watcher of the ways of nature; she never shows her operations to a man in a bustle. Patience, perseverance, anxious but undemonstrated solicitude, assiduity amounting to zeal, but never showing itself in hurry or fretfulness, these are the qualities that win access to her most secret wonders. . . . With a punctuality and accuracy worthy of all imitation, [White] registered such changes and appearances as contributed to the general stock of information; exaggeration was unknown to him, for he had no purpose of effect to answer, was careless of fame, but so true a lover of his pursuit, that he would have held any deviation from exactness as something shocking and revolting. Men of pure and simple minds are usually remarkable for accuracy. . . . The letters now comprised in one three shilling volume, are the result of the observations of a long life. The phenomena he describes, are not so curious in themselves,—though there are a multitude of new remarks, and many of them have become the germs of subsequent knowledge,—as the spirit in which they are described is graceful and engaging. This is the charm of the work; it is this which has made it the favourite of every one in whose way it has fallen. A philosopher whose mind was never disturbed, whose memory was well stored, whose taste was pure and simple, and who was perfect master of his subject, naturally expressed himself in an elegant style. He had too a gentle fancy, which combined with a benevolent complaisance, lent a sort of glow to his composition. His sentences flow as placidly as did the days of his life, and like them, each has its little incident, its remark or inference interesting in itself, lively in the manner of conception, and agreeable in the mode of communication. His registers and his letters were, in fact, the mere reflection of his life.

The study of Natural History has made great progress of late; and much of the impetus it has received is to be attributed to the History of Selborne.

'White's *Selborne*', *Westminster Review*, 19 (July 1833), 39–42, at 40, 41–2.

White's Natural History of Selborne is a most popular little book, and deservedly so, though Selborne itself be but an obscure parish somewhere in the south of England. The very local title of the work has not in the least militated against its interest. But why? Partly, it would seem, from the very pleasing manner in which it is written; partly because the natural history of even a single parish may be regarded as that of the entire district in which the parish is included. And may not the germ of a similar popularity be found, if the writer not fail in his part, in the traditional history of a Scotch village?

. . . What does it matter that we examine only a little part of any substance, if from that part we acquire an ability to judge of the whole? The chemist can subject but comparatively small portions of his earths or metals to the text of experiment, but of how wide an application are the laws which he discovers in the process!

> Scottish geologist, stonemason, and journalist Hugh Miller, in his dedication to the geologist Thomas Dick Lauder, in *Scenes and Legends of the North of Scotland, or The Traditional History of Cromarty* (Edinburgh, 1835), pp. vii, viii.

The spread of education has improved the intellectual taste of the public, which has grown so critical, that nothing short of high merit will please. In this state of affairs, we in all kindness would recommend our poetically-inclined friends to turn their mental energies to better account than hammering crude ideas into verses. There is scarcely a district of country which does not offer something worthy of noting down and describing, be it even for private recreation and literary discipline. The 'Natural History of Selborne,' one of the most pleasing books that was ever published, is exactly of this nature. Now, it is in the power of almost *every* person to write such a book, though not so cleverly and poetically, perhaps, as the

Rev. Gilbert White. Would, therefore, our amateur-poets favour us with works of this class, or the printed results of any branch of useful investigation in sober and sensible prose, we shall not only feel grateful, but do all in our power to advance their views: they would also advance their own; for, having stored up a fund of knowledge, their imaginations would take a healthy and vigorous tone, their poetical faculties would expand and brighten, and they would become poets in the best signification of that much-abused word.

> 'Amateur Poets', *Chambers's Edinburgh Journal* (20 May 1843), 141–2 (probably written by the publisher and scientific writer Robert Chambers).

As for Thoreau, there is one chance in a thousand that he might write a most excellent and readable book. . . . He is the most unmalleable fellow alive—the most tedious, tiresome, and intolerable—the narrowest and most notional—and yet, true as all this is, he has great qualities of intellect and character. The only way, however, in which he could ever approach the popular mind, would be by writing a book of simple observation of nature, somewhat in the vein of White's History of Selborne.

> American novelist Nathaniel Hawthorne, letter to the New York publisher E. A. Duyckinck, 1 July 1845, *Nathaniel Hawthorne: The Letters, 1843–1853*, ed. Thomas Woodson, L. Neal Smith, and Norman Holmes Pearson (Columbus, Ohio, 1985), 106.

Let the world also share in your beautiful communion with Nature, with the *mute* Tribes of the Living! Nothing is more universally interesting; as indeed it may well be: a silent practical worship and recognition of this universe; most welcome in these loud days of mouth-worship and sham-recognition! Do you know a little English Book we have, called White's *Natural History of Selborne*? It should be in the Duke's Library: if it be not accessible, tell me, and I will get it sent to you. You should do something like it for your own corner of Germany; you really should. It is one of our most excellent Books; White, a quiet country Parson, has preached a better sermon here than all the loud Bishops that then were. . . .

I have a reasonable little Copy of White's *Nat^l History of Selborne*; which I will send . . . . Of all the Books of Natural History I have ever read, White's, in his healthy genial limitation, pleases me most. Go and do likewise!

> Historian Thomas Carlyle, letters to Johann Peter Eckermann, 2 August 1848 and 5 January 1849, *The Carlyle Letters Online*, 2007,<http://carlyleletters.org> (accessed 7 January 2013).

I may add that I am an observer of nature generally, and the character of my observations, so far as they are scientific, may be inferred from the fact that I am especially attracted by such books of science as White's Selborne and Humboldt's 'Aspects of Nature.'

> American transcendentalist and nature writer Henry David Thoreau, response to a questionnaire that accompanied his refusal in December 1853 of membership of the Association for the Advancement of Science, *The Correspondence of Henry David Thoreau*, ed. Walter Harding and Carl Bode (New York, 1958), 309–10.

Since White's 'History of Selborne,' few or no writers on Natural History, save Mr. [Philip Henry] Gosse and poor Mr. [Edward] Forbes, have had the power of bringing out the human side of science, and giving to seemingly dry disquisitions and animals of the lowest type, by little touches of pathos and humour, that living and personal interest, to bestow which is generally the special function of the poet . . .

> Clergyman, novelist and naturalist Charles Kingsley, in *Glaucus; or, The Wonders of the Shore* (Cambridge, 1855), 160.

Headache. Read Lucrezia Floriani. We are reading White's History of Selborne in the evening . . .

> Novelist and essayist, George Eliot, in her journal entry for 22–3 June 1855, *The Journals of George Eliot*, ed. Margaret Harris (Cambridge, 1998), 57.

It was the pages of Gilbert White and the woodcuts of Bewick which first beguiled the English schoolboy to the observation of our feathered friends. . . . Few men have ever attained eminence in science whose minds were not early attracted to the subject; and when style has happily combined with truth and nature to rivet the attention of childhood, no slight service has been rendered to the cause of natural history. Vast as has been the advance in systematic knowledge within the last half-century, how few of our living naturalists but must gratefully acknowledge their debt to White's 'History' and to the life-like woodcuts of Bewick! Probably we shall not wrong the cultivated annalist of Selborne by giving the first place to Bewick. We are tempted to believe that for one studious schoolboy whose latent taste has been evoked by the former, a dozen have been led 'how to observe' by conning over 'The British Birds' on a holiday afternoon in their father's library. Yet Bewick has not the slightest claim to rank with Gilbert White as a naturalist. White was what Bewick never was, a man of science; but, if no naturalist, Bewick was a lover of nature, a careful observer, and a faithful copier of her ever-varying forms. In this, and this alone, lies his charm.

'Review of the recently published memoir of Bewick', *Ibis: A Magazine of General Ornithology* 4 (October 1862): 368–78, at 368.

I have never thanked you for your work on the Orchidiæ—which I read with wonder & admiration. It confirms my faith that White of Selborne bequeathes his mantle to you—

American physician and botanist Francis Boott, resident in London, in a letter to Charles Darwin, 22 December [1862], *The Correspondence of Charles Darwin*, ed. F. H. Burckhardt et al. (Cambridge, 1985–), x, 621.

WINDBAG:

Yet one last name, O Birds, I will engage
In our behalf. Remember Selborne's sage!
How oft have ye beheld his footsteps crown
Your leafy hanger, or your open down,
Or strayed sometimes where the dwarf oak-tree veins
With crooked and sprawling roots your sandstone lanes!
He, bright historian of your loves and feuds,
Dated your building, chronicled your broods,
Described your times of flight, your change of feathers,
Your light moods shifted with the shifting weathers,
And, by long commerce with his gable guests,
Learned all the secrets of your souls and nests.
Chiefly to you I plead, whose airy host,
And manifold migrations, pleased him most.
How strained his eye for that first April comer,
Promise of old companions and the summer!
How pensive watched your 'placid easy flight'
Southwards! Ye swallows, think of Gilbert White! . . .

SWALLOW:

Books he shall read in hill and tree;
The flowers his weather shall portend;
The birds his moralists shall be;
And everything his friend.

Such man in England I have seen;
He moved my heart with fresh delight;
And had I not the swallow been,
I had been Gilbert White.

Poet William John Courthope, in the voices of his character Windbag, a poet of the Romantic school, and Swallow, a bird in paradise, *The Paradise of Birds* (Edinburgh, 1870), 123–5, 126–7.

Open the book where you will, it takes you out of doors. In our broiling July weather one can walk out with this genially garrulous Fellow of Oriel and find refreshment instead of fatigue. You have no trouble in keeping abreast of him as he ambles along on his hobby-horse, now pointing to a pretty view, now stopping to watch the motions of a bird or an insect, or to bag a specimen . . . The book has also the delightfulness of absolute leisure. Mr. White seems never to have had any harder work to do than to study the habits of his feathered fellow-townsfolk, or to watch the ripening of his peaches on the wall. His volumes are the journal of Adam in Paradise . . . .

It is positive rest only to look into that garden of his . . . here the world has no entrance. No rumor of the revolt of the American Colonies seems to have reached him. 'The natural term of a hog's life' has more interest for him than that of an empire. . . .

Another secret charm of this book is its inadvertent humor, so much the more delicious because unsuspected by the author. How pleasant is his innocent vanity in adding to the list of the British, and still more of the Selbornian, *fauna*! I believe he would gladly have consented to be eaten by a tiger or a crocodile, if by that means the occasional presence within the parish limits of either of these anthropophagous brutes could have been established. He brags of no fine society, but is plainly a little elated by 'having considerable acquaintance with a tame brown owl.' Most of us have known our share of owls, but few can boast of intimacy with a feathered one. . . .

In 1770 he makes the acquaintance in Sussex of 'an old family tortoise,' which had then been domesticated for thirty years. It is clear that he fell in love with it at first sight. We have no means of tracing the growth of his passion; but in 1780 we find him eloping with its object in a post-chaise. 'The rattle and hurry of the journey so perfectly roused it that, when I turned it out in a border, it walked twice down to the bottom of my garden.' It reads like a Court Journal: 'Yesterday morning H. R. H. the Princess Alice took an airing of half an hour on the terrace of Windsor Castle.' This tortoise might have been a member of the Royal Society, if he could have condescended to so ignoble an ambition. It had but just been discovered that a surface inclined at a certain angle with the plane of the horizon took more of the sun's rays. The tortoise had always known this (though he unostentatiously made no parade of it), and used accordingly to tilt himself up against the garden-wall in the autumn. He seems to have been more of a philosopher than even Mr. White himself, caring for nothing but to get under a cabbage-leaf when it rained, or the sun was too hot, and to bury himself alive before frost,—a four-footed Diogenes, who carried his tub on his back.

There are moods in which this kind of history is infinitely refreshing. These creatures whom we affect to look down upon as the drudges of instinct are members of a commonwealth whose constitution rests on immovable bases. Never any need of reconstruction there! *They* never dream of settling it by vote that eight hours are equal to ten, or that one creature is as clever as another and no more. . . . It is good for us now and then to converse with a world like Mr. White's, where Man is the least important of animals. But one who, like me, has always lived in the country and always on the same spot, is drawn to his book by other occult sympathies. Do we not share his indignation at that stupid Martin who had graduated his thermometer no lower than 4° above zero of Fahrenheit, so that in the coldest weather ever known the mercury basely absconded into the bulb, and left us to see the victory slip through our fingers just as they were closing upon it? No man, I suspect, ever lived long in the country without being bitten by these meteorological ambitions. He likes to be hotter and colder, to have been more deeply snowed up, to have more trees and larger blown down than his neighbors. With us descendants of the Puritans especially, these weather-competitions supply the abnegated excitement of the race-course. . . . I seem to glimpse something of this familiar weakness in Mr. White. He, too, has shared in these mercurial triumphs and defeats.

American writer and critic James Russell Lowell, 'My Garden Acquaintance', in *My Study Windows* (Boston, 1871), 1–23, at 1–5.

. . . we have no natural history of birds written yet. It cannot be written but by a scholar and a gentleman; and no English gentleman in recent times has ever thought of birds except as flying targets, or flavourous dishes. The only piece of natural history worth the name in the English language, that I know of, is in the few lines of Milton on the Creation. The only example of a proper manner of contribution to natural history is in White's Letters from Selborne.

Art critic John Ruskin, in *Love's Meinie: Lectures on Greek and English Birds, Given before the University of Oxford*, 3 vols. (Keston, Kent, 1873–81), i, 7–8.

Messrs. Macmillan having requested me to edit White's 'Selborne,' I accepted the task, feeling assured that the handsome Edition of the works of the founder and pioneer of English Practical Natural History now presented to the public would be the means of attracting many of the present generation—both young and old—to the observation of the living works of the great Creator, and would help to counteract the growth of doubt, infidelity, and atheism, which—though regarded at their real worth by a reasoning public—must become bitter weeds in future, of no assistance to science, and sure promoters of a dangerous materialism. . . .

White was a true student of all created things—lynx-eyed, quick to observe accurately, and patient to interpret the meaning of facts brought under his notice. The same facts that White saw and recorded are still going on around us at the present time. The birds come and go at the same dates as did their ancestors a century ago. The rabbits, hedgehogs, rats, bats, snakes, mice, &c., still keep up their old, old customs unaltered and unchanged. White is the teacher who has shown four generations how and what to observe—in fact, he taught them the 'Art of Observation.' For the above reason, therefore, White's 'Selborne' has held its own as a standard book for a hundred years, and will probably be as fresh as ever a hundred years hence. . . .

White's 'Selborne' again will show clergymen that they have many parishioners inhabiting the woodlands, hedges, and fields, whose welfare they would do well not to neglect. There is hardly a parish in England or Wales where the clergyman has not opportunities more or less favourable for writing a local 'White's Selborne,' taking White's method of observing and recording as a model for his note-book.

I feel assured that the education of children, both in town and country, might greatly be forwarded if they were taught in the schools what and how to observe. Especially in the country should they be encouraged to make collections of common objects, animal, vegetable, and mineral. They should also be taught to recognise indigenous British birds and beasts, and to send in notes as to what they have observed of their habits. Such studies tend to sharpen the natural faculties, while they humanize the intellect.

> Naturalist and fish specialist Frank Buckland, in the Preface to his edition of *The Natural History of Selborne* (London, 1875), pp. vii–xii.

There is a singular parallel in the popularity of the two old books, the 'Complete Angler' of Isaac Walton, and the 'Natural History of Selborne,' by the Rev. Gilbert White. This popularity has gone on steadily increasing in both cases, until both books are of that class which everyone has read or is supposed to have read, or, with reference to the coming generation of readers, ought to read. The cause of the esteem in which the two books are held is mainly the same. Honest, manly, and godly in their tone, simple and clear in their style, with no ostentation, clearness and accuracy of observation in those subjects which each particularly affected, and with the charm of enthusiasm, and enthusiasm with respect to the glorious 'out of doors,' they are models for all succeeding writers on kindred subjects.

> George Christopher Davies, conservator of fisheries of Norfolk and Suffolk, and writer on country pursuits, in the Introduction to his edition of *The Natural History of Selborne* (London and New York, 1879), p. ix.

When the writer of the little volume now offered to the Canadian public first settled in the then unbroken back-woods, on the borders of the Katchawanook, just where the upper waters of a chain of lakes narrow into rapids of the wildly beautiful Otonabee; the country at that time was an unbroken wilderness. . . .

Having no resource in botanical works . . . I relied entirely upon my own powers of observation, and this did very much to enhance my interest in my adopted country and add to my pleasure as a relief, at times, from the home longings that always arise in the heart of the exile, especially when the sweet opening days of Spring return to the memory of the immigrant . . . But for the Canadian forest flowers, and trees and shrubs, and the lovely ferns and mosses, I think I should not have been as contented as I have been, away from dear old England. It was in the hope of leading other lonely hearts to enjoy the same pleasant recreation, that I have so often pointed out the natural beauties of this country to their attention, and now present my forest gleanings to them in a simple form, trusting that it may not prove an unacceptable addition to the literature of Canada, and that it may become a household book, as Gilbert White's Natural History of Selborne is to this day among English readers . . .

> Canadian poet and naturalist Catharine Parr Traill, writing in 1884 from Lakefield, Ontario, in *Studies of Plant Life in Canada* (Ottawa, 1885), 1, 3.

The simple character of the writings of Gilbert White have, perhaps, in these latter days somewhat deterred people from reading him. They seem so very simple, just as a boy might snatch a bloom of horse-chestnut and bring it home; so very easy to do that. And what is there in a chestnut bough when you have got it? In these times there is a certain amount of feeling abroad that all books of information must be written with a marked hardness of style. They must abound in long words and sentences that come out with a slow, crushed motion, like a rail from the rolling-mill. More particularly is this the case with books of what is called popularised knowledge—accurate, comprehensive, fitted in and square, like the mosaic tiles now so common in grates, and about as charming. They contain everything; they tell you in every page they are teaching you, and leave you without a single idea fixed in the memory. If any one accustomed to such heavily-edited volumes picks up the *Natural History of Selborne* it is almost like a shock. The sense of repose is at first taken for lack of ability—the author is not one of the 'ablest' writers of the day; very inferior! . . .

A great master is under a disadvantage. You go to look at an old and celebrated picture with exalted feelings, and when you get there you say, 'How disappointing! I have seen all this before; the style, the attitude, and the method of composition are all familiar in a hundred engravings and modern pictures, and, really, the old masters, instead of being such a guide, look a long way behind the age. We can do things better now.' The secret is, the old master's work has been multiplied exceedingly, and used as the ground-work on which to build innumerable variations. Without his work these could never have come into existence. From the stores accumulated by Gilbert White a very great deal of the contents of modern books have been drawn. Not only the facts but the general system has been followed out in a hundred ways, so that his book suffers exactly like the old picture, until you understand it. The more you understand them the more you appreciate old masters, whether artists or authors, until you would be ready, if you had the means, to give the extraordinary prices for them that seem so incomprehensible to outsiders.

Writer and mystic Richard Jefferies, in the Preface to his edition of *The Natural History of Selborne* (London, 1887), pp. vii, x.

I was moved to take down my White's 'Selborne' and examine it again for the source of the delight I had had in it, on hearing a distinguished literary man . . . say it was a book he could not read with any degree of pleasure: to him its pages were a bare record of uninteresting facts. . . . The book was doubtless pitched in too low a key for him: it was tame and commonplace, like the country itself. There is indeed something a little disappointing in White's book when one takes it up for the first time, with his mind full of its great fame. It is not seasoned quite up to the modern taste. White is content that the facts of nature should be just what they are; his concern alone is to see them just as they are. When I myself first looked into his book, many years ago, I found nothing in it that attracted me, and so passed it by. Much more recently it fell into my hands, when I felt its charm and value at once. Indeed, the work of the Selborne naturalist belongs to the class of books that one must discover for himself: their quality is not patent; he that runs may not read them. . . . Its brevity, its directness, its simplicity, its dealings with familiar and near-at-hand objects, shows, occurrences, etc., make it a book which never sates and never tires the reader. It is little more than an appetizer, but as such it takes high rank. As a stimulus and spur to the study of natural history, it has no doubt had more influence than any other work of the century. Its merits in this direction alone would perhaps account for its success. But, while it has other merits, and great ones, it has been a fortunate book: it has had little competition; it has had the wind always with it, so to speak. It furnished a staple the demand for which was always steady and the supply small. There was no other book of any merit like it for nearly a hundred years. . . . Hence we say that while it is not one of the great books, it is one of the very *real* books, one of the very *live* books, and has met and supplied a tangible want in the English reading world. It does not appeal to a large class of readers, and yet no library is complete without it. It is valuable as a storehouse of facts, it is valuable as a treatise on the art of observing things, and it is valuable for its sweetness and charm of style. . . . There is no evidence that [White] was anything like the petulant recluse and man-hater that our own Thoreau at times was. He had the wide, generous eye, and his love of nature was not in any sense a running away from the world.

American nature writer John Burroughs, in his essay 'Gilbert White's Book', *Indoor Studies* (Boston, 1889), 162–76, at 162–6.

SELBORNE

How oft in sickness when the languid brain
    Longed for the freshness of a summer wood,
And the tired reason could not bear the strain
    Of ordered thinking which before it stood,
Have I, so longing, just re-read the page
    Of him who wrote of Selborne and its birds,
To whom through years of slow and peaceful age
    Did kindly Nature whisper all her words . . .

And so I turned and turned again the leaves
    Of that old record of a charmèd house,
Saw the weird bats fast dropping from the eaves,
    Or climbed the wheat-stalks with the 'Harvest Mouse.'
Until my wearied eyes were closed in sleep,
    And woke refreshed as when the night is done,
Saw with fresh health the morning sunbeams sweep,
    And heard at dusk the Plymouth evening gun.
And so in every English-speaking land
    The name of this dear Selborne stands in light.
For Nature's voice and all her bounteous hand
    Still sings and plays to men through Gilbert White.

Politician and anti-Darwinian George Douglas Campbell, 8th Duke of Argyll, lines from the first and last stanzas of his poem 'Selborne', *Burdens of Belief and other Poems* (London, 1894), 82, 84–5.

So many learned and elaborate treatises have sunk beneath the waves upon which this cockle-shell of a book rides so safely and buoyantly! What is the secret of its longevity? One can do little more than name its qualities without tracing them to their sources. It is simple and wholesome, like bread, or meat, or milk. Perhaps it is just this same unstrained quality that keeps the book alive. . . .

The style of the born writer is like an open fire: we are in direct communication with his mind; we see the play of the forces at work; we get that precious sense of reality. All this is true of White's pages. Yet he had no literary ambitions. His style is that of a scholar, but of a scholar devoted to natural knowledge. There was evidently something winsome and charming about the man personally, and these qualities reappear in his pages.

White was a type of the true observer—the man with the detective eye. He did not seek to read his own thoughts and theories into Nature, but submitted his mind to her with absolute frankness and ingenuousness. He had infinite curiosity, and delighted in nothing so much as a new fact about the birds and the wild life around him. To see the thing as it was in itself and in its relations, that was his ambition.

American nature writer John Burroughs, in the Preface to his edition of *The Natural History of Selborne* (New York, 1895), pp. vii, xi.

White still keeps his hold upon the English mind. Though many may talk of his book without having read it through, and still more may have read it through without discerning its quality, the fact remains that the demand for it is steadily on the increase, and that it is finding its way into the village library and the home of the working man.

Ornithologist and historian William Warde Fowler, in *Summer Studies of Birds and Books* (London, 1895), 203.

It was the business of White's generation to substitute careful and accurate first-hand observation for the vague descriptions, the false surmises, and the wild traditional tales of earlier authors.

This it is in great part that gives their perennial charm to these natural, personal, and delightful Letters. We are present, as it were, at the birth of zoology; we are admitted to see science in the making. . . . It is therefore as a historic document that the 'Natural History of Selborne' most of all appeals to us; it shows us by what steps science felt its way in the later years of the eighteenth century. . . .

Book-knowledge tends more and more to supersede direct contact with nature. But White may suggest to us a more excellent way. The record of his long years spent in finding out for himself what the beasts and birds really did do makes us feel that books are of little use beside direct eyesight. . . .

Indeed, throughout, White was one of the few early naturalists who recognised the importance of the cumulative effect of infinitesimal factors—a truth on which almost the whole of modern biology and geology are built up. As zoologist, as botanist, as meteorologist, as sociologist, he is possessed in anticipation by the modern spirit in every direction. . . .

Finally, we have always to bear in mind the end which thinkers of White's age proposed to themselves. In our own day, the desire to 'advance science' has been made on the whole a foolish fetish. Almost all scientific education has aimed at this end; it has striven to produce, not whole and many-sided men and women, but inventors, discoverers, producers of new chemical compounds, investigators of new and petty peculiarities in the economy of the greenfly that affects roses. All that is very excellent in its way; but it is not the sole, or even (let me be frank) the main object of a scientific education. What the world needs is not so much advancers of science as a vast mass of well-instructed citizens, who can judge of all subjects alike in their proper place, and can assign to each its due relative importance. . . . The truth is, the vast majority of men can never do anything to 'advance science' in any noteworthy degree . . . The aim we should propose is to build ourselves up in the round . . . As a preservative against one prevalent form of lop-sidedness in modern life, White's methods and example are of incalculable value.

> Science writer and novelist, Grant Allen, in the Introduction to his edition of *The Natural History of Selborne* (London, 1900), pp. xxxi–xxxix.

That White's 'Selborne' is the only work on natural history which has attained the rank of an English classic is admitted by general acclamation, as well as by competent critics, and numerous have been the attempts to discover the secret of its ever-growing reputation. Scarcely two of them agree, and no explanation whatever offered of the charm which invests it can be accepted as in itself satisfactory. If we grant what is partially true, that it was the first book of its kind to appear in this country, and therefore had no rivals to encounter before its reputation was established, we find that alone insufficient to account for the way in which it is still welcomed by thousands of readers, to many of whom—and this especially applies to its American admirers—scarcely a plant or an animal mentioned in it is familiar, or even known but by name. . . .

The secret of the charm must be sought elsewhere; but it has been sought in vain. Some have ascribed it to his way of identifying himself in feeling with the animal kingdom, though to this sympathy there were notable exceptions. Some, like Lowell, set down the 'natural magic' of White to the fact that, 'open the book where you will, it takes you out of doors;' but the same is to be said of other writers who yet remain comparatively undistinguished. White's style, a certain stiffness characteristic of the period being admitted, is eminently unaffected, even when he is 'didactic,' as he more than once apologises for becoming, and the same simplicity is observable in his letters to members of his family, which could never have been penned with the view of publication, and have never been retouched. Then, too, there is the complete absence of self-importance or self-consciousness. The observation or the remark stands on its own merit, and gains nothing because he happens to be the maker of it, except it be in the tinge of humour that often delicately pervades it. The beauties of the work, apart from the way in which they directly appeal to naturalists, as they did to Darwin, grow upon the reader who is not a naturalist, as Lowell testifies, and the more they are studied the more they seem to defeat analysis.

> Zoologist Alfred Newton, in his article on Gilbert White in the *Dictionary of National Biography* (1900).

[S]ome modern editors of the famous 'Letters' . . . are astonished at Gilbert White's vitality, and cannot find a reason for it. Why does this 'little cockle-shell of a book,' as one of them has lately called it, come gaily down to us over a sea full of waves, where so many brave barks have foundered? The style is sweet and clear, but a book cannot live merely because it is well written. It is chock-full of facts; but the facts have been tested and sifted, and all that were worth keeping are to be found incorporated in scores of standard works on natural history. I would humbly suggest that there is no mystery at all about it; that the personality of the author is the principal charm of the 'Letters,' for in spite of his modesty and extreme reticence his spirit shines in every page; that the world will not willingly let this small book die, not only because it is small, and well written, and full of interesting matter, but chiefly because it is a very delightful human document.

> Naturalist and nature writer William Henry Hudson, in *Birds and Man* (London, 1901), 310.

No explanation of the merit of the *Natural History of Selborne* can be at all adequate which does not dwell upon the reality and truth of White's descriptions. His personal knowledge of nature was great, not in relation to the knowledge accumulated in books, but in comparison with the direct experience of most other naturalists of any age. Here is one great difference between him and the imitators who have hoped to succeed by mere picturesque writing. White is interesting because nature is interesting; his

descriptions are founded upon natural fact, exactly observed and saga-
ciously interpreted. Very few of his observations and not many of his infer-
ences need correction more than a hundred years after his death. . . .

Probably no book in any language has incited more people to take up the
study of natural history. Many have tried to write letters on White's plan,
but their eyes were not so good as his; they failed to pick out the points of
real interest, and they had no Selbornes to describe. By the use of natural
gifts exactly suited to the task a man who was sagacious rather than profound, and well-read rather than learned, wrote us a book which will
endure for ever, a small thing perfectly done.

> Naturalist Louis Compton Miall, in the Introduction to his edition, with
> W. Warde Fowler, of *The Natural History of Selborne* (London, 1901),
> pp. xxii, xxv–xxvi.

It is not always the most confidently conducted books, or those best preceded by blasts on the public trumpet, which are eventually received with
highest honours into the palace of literature. No more curious incident of
this fact is to be found than is presented by the personal history of that
enchanting classic, White's *Selborne*. If ever an author hesitated and
reflected, dipped his toe into the bath of publicity, and hastily withdrew it
again, loitered on the brink and could not be induced to plunge, it was the
Rev. Gilbert White. This man of singular genius was not to be persuaded
that the town would tolerate his lucubrations. He was ready to make a
present of them to any one who would father them, he allowed his life to
slip by until his seventieth year was reached, before he would print them,
and when they appeared, he could not find the courage to put his name on
the title-page. Not one of his own titlarks or sedge-warblers could be more
shy of public observation. Even the fact that his own brother was a publisher gave him no real confidence in printers' ink. . . .

Countless as have been the reprints of *The Natural History of Selborne*,
its original form is no longer, perhaps, familiar to many readers. The first
edition, which is now before me, is a very handsome quarto. Benjamin
White, the publisher, who was the younger brother of Gilbert, issued most
of the standard works on natural history which appeared in London during
the second half of the century, and his experience enabled him to do adequate justice to *The History of Selborne*.

> Critic and biographer Edmund Gosse, in his essay 'The Natural History of
> Selborne', *Gossip in a Library* (London, 1913), 195, 198.

[*In Nature's Ways*] was planned and written with the purpose of introducing young people to Gilbert White's 'Natural History of Selborne,' and to
encourage the study of that immortal work. . . .

[*Selborne*] is not suitable for young readers, or readers young in the study of the natural history of an English countryside. The quaint and sometimes magisterial air of some of the letters, their geography and their geology, make them hard reading; the book is like a country ringed about by rugged mountain barriers which keep many travellers from the delectable valleys of the interior. Yet how full of delight for all young lovers of Nature, how full of the sense of 'everything that is rural, verdurous, and joyous,' like the field-cricket's song, are selected passages! . . .

I have chosen the passages on the general plan of giving those which express simple truths, so that any teacher may take them at random, and be sure that they will serve for lessons, a word which in such a case will lose its usual meaning to children. I have avoided passages where the information given, or the theory advanced, is faulty in the light of newer knowledge, but this only as a general rule; for those who understand Gilbert White and his writings and his times find no fault with his faults, and feel that much of the charm of the letters lies in his doubts, questionings, tentative theories, half believed in and half disbelieved, his groping in the darkness of his age after the truth—by which he teaches us how to deduct and reason.

Should Fates prove kindly, I hope to be able to publish a second and more advanced volume, including all the now neglected pages in the 'History' which fall into the general plan of the work—to be an introduction (if I may emphasise my purpose) to one of the greatest books in the whole world.

> Nature writer Marcus Woodward, in his book *In Nature's Ways: A Book for all Young Lovers of Nature, Being an Introduction to Gilbert White's 'Natural History of Selborne'* (London, 1914), 5–6.

He had, in fact, made a book which had three extraordinary merits. It contained valuable and new observations; it overflowed with evidence of a new spirit—a spirit of minute and even loving inquiry into the life and personality of animals in their native surroundings—that was coming into natural history; and, thirdly, it had style or whatever we like to call the breath of life in written words, and it was delightfully and easily full of the man himself and of the delicate eighteenth-century southern countryside which he knew. But the observations are no longer new; the new spirit has been renovated by the gunless naturalists from Thoreau to Mr W. H. Hudson of our own day. The man himself is still fresh to succeeding generations, and thousands, who care not at all how many willow-wrens there be, delight to read these letters from a man so happy and remote from our time that he thought the dying fall of the true willow-wren 'a joyous easy laughing note.' We are always pleasantly conscious of the man in his style, which strikes us as the lines and motions of a person's face strike us for good or for bad, and, even so, in a manner that defies analysis. His quack who ate a toad, his boys twisting the nests out of rabbit-holes with a forked stick, his love of the 'shapely-figured aspect of the chalk hills' above that of the 'abrupt and shapeless' mountains, his swallows feeding their young and treading on the wing, his friendly horse and fowl, his prodigious many-littered half-bred bantam sow that proved, 'when fat, good bacon, juicy and tender,' his honey-loving idiot, his crickets ('a good Christmas fire is to them like the heats of the dog-days')—these things have in his pages a value which can only be attributed to his literary genius, by which his book survives.

> Critic, war poet, and writer of nature books Edward Thomas, in *A Literary Pilgrim in England* (London, 1917), 109, 114–15.

Gilbert White discovered the formula for complete happiness, but he died before making the announcement, leaving it for me to do so. It is to be very busy with the unimportant. The author of *The Natural History of Selborne* became so interested in the birds and bats, the frogs and fishes, even the vermin of his parish, for he was a clergyman by trade, that he entirely forgot everything that was going on in the world—he even forgot to get married. I love an absent-minded man . . . .

> American author and publisher Alfred Edward Newton, in *This Book-Collecting Game* (Boston, 1928), 266.

Selborne . . . has the fine compression and vitality which can only be given to facts after long study and speculation by a vigorous, original mind. It has, at the same time, a notable restraint; the author is on his guard against betraying the enthusiasm he frequently feels. . . . This restraint has much to do with the success of the Natural History. It is a work of art, not all of the same merit but all in the same pattern. It includes nothing to break the spell of the imaginary world it creates. For Gilbert White's world is imaginary. It is not Nature. It is the finest and broadest and the most enduring bridge that can be built between Nature and civilization; it has elements of both, but it is in the last resort true to neither. It is perhaps the nearest approach to Nature which can be universally understood; it opens a world which is not a human world, but still is ordered where Nature is unkempt, polished and serene, where Nature is ebullient, and individually aware where Nature is collectively blind. *Selborne* is commonly taken to be an artless and spontaneous composition; actually it was shaped with as much labour as any other work of art of equal importance.

Ornithologist, environmentalist, and advocate for nature conservancy (Edward) Max Nicholson (www.maxnicholson.com), in the Introduction to his edition of *The Natural History of Selborne* (London, 1929), 34–5.

I began now to borrow from the Sanatorium Library books on nature and the countryside—Hardy, Hudson, Jefferies, Gilbert White; books on birds, animals, snakes and trees. . . . And all these presented a picture of an England which, except in a few secluded spots, no longer survived.

Norman Nicholson, a shopkeeper's son, writing between 1 June 1932 and 31 December 1932, when an inmate of Linford Sanatorium, Hampshire, *Wednesday Early Closing* (London, 1975), 190.

With the exception of Shakespeare, Milton and Wordsworth, there can hardly be another English man of letters to whom a larger body of comment and criticism has been devoted than Gilbert White. . . . What Thomas Aquinas was to the Church, Plutarch to biography, Chaucer to the mediaeval English spirit, Tennyson to the dreams of the nineteenth century, Gilbert White was and remains not only to naturalists but to every student of a countryside which has only recently ceased to be the backbone of England. He is and always has been the Saturn of our green Olympiad . . . .

It was only when I came to reread a text which has yearly shrunk in proportion to the seas of marginalia that have broken upon it, that, as is perhaps true of the work of all great men, *aliquid novi* occurred to me as still to be extracted from this *terra incognita*. What made a particular impression upon me was the lack of attachment in the father of natural history not only to the age into which he was born, but to our present time, with which he is said to have more affinities than with his own. Where then does he historically belong? . . .

As a stylist, Bunyan always wore worsted, but the parson-squire of Selborne wore gaiters like any other gaffer. It was this contact with village, field and woodland not merely by choice of mind but in their habitual effect upon his self-expression which isolates Gilbert White's style from the norm of that of his contemporaries, while it accepts without question the current phraseology. . . .

Critics of White the naturalist love to put him ahead of his own time, and some of the more progressive modern ones to modernize him right out of his own skin. Speaking of him as a literary craftsman, I should be inclined not to strap him upon the Time Machine at all, but to attach him to place rather than to time. . . . The life of locality was what infused into his literary content a something more than eloquent and scholarly English. . . .

When it came to the actual substance of what he wrote, Gilbert White took as abrupt and final a leave of his contemporaries as Robinson Crusoe. He went off to what, so far as the eighteenth century wits, moralists, scholars and poets were concerned, was a desert island. . . .

Just as Selborne itself represents an individual piece of the English countryside, so its historian is a perfect example of the interdependence between the local and the universal, the particular and the general. Gilbert White is the supreme voice of the English parish, and it is this, his most important quality, that attaches him to the Middle Ages rather than to any other phase of our historical past.

> Rural writer Harold John Massingham, in the Introduction to his edition
> of *The Writings of Gilbert White of Selborne* (London, 1938), at pp. ix,
> xi–xii, xv, xxiv.

It is one of those ambiguous books that seem to tell a plain story, the *Natural History of Selborne*, and yet by some apparently unconscious device of the author's has a door left open, through which we hear distant sounds, a dog barking, cart wheels creaking, and see, when 'all the fading landscape sinks in night', if not Venus herself at least a phantom owl.

His intention seems plain enough—it was to impart certain observations upon the fauna and flora of his native village to his friends Thomas Pennant and Daines Barrington. But it was not for the benefit of those gentlemen that he composed the sober yet stately description of Selborne with which the book opens. . . . No novelist could have opened better. Selborne is set solidly in the foreground. But something is lacking; and so before the scene fills with birds, mice, voles, crickets and the Duke of Richmond's moose, before the page is loud with the chirpings, bleatings, lowings and gruntings of their familiar intercourse, we have Queen Anne lying on the bank to watch the deer driven past. It was an anecdote, he casually remarks, that he had from an old keeper, Adams, whose great-grandfather, father and self were all keepers in the forest. And thus the single straggling street is allied with history, and shaded by tradition. No novelist could have given us more briefly and completely all that we need to know before the story begins.

The story of Selborne is a vegetable, an animal story. The gossip is about the habits of vipers and the love interest is supplied chiefly by frogs. Compared with Gilbert White the most realistic of novelists is a rash romantic. The crop of the cuckoo is examined; the viper is dissected; the grasshopper is sought with a pliant grass blade in its hole; the mouse is measured and found to weigh one copper halfpenny. Nothing can exceed the minuteness of these observations, or the scrupulous care with which they are conducted. The chief question in dispute—it is indeed the theme of the book—is the migration of swallows. . . . Every grain of evidence is sifted; none is obscured. With all his faculties bent on this great question, the image of science at her most innocent and most sincere, he loses that self-consciousness which so often separates us from our fellow-creatures and becomes like a bird seen through a field-glass busy in a distant hedge. This is the moment then, when his eyes are fixed upon the swallow, to watch Gilbert White himself. . . .

And just as the vicarage garden seemed to Aunt Snookes tortoise a whole world, so as we look through the eyes of Gilbert White England becomes immense. The South Downs, across which he rides year after year, turn to 'a vast range of mountains.' The country is very empty. He is more solitary at Selborne than a peasant to-day in the remotest Hebrides. . . . unconscious happiness is conveyed, not by assertion, but much more effectively by those unsought memories that come of their own accord unsought. They are all of hot summer evenings—at Oxford in Christ Church quadrangle; riding from Richmond to Sunbury with the swallows skimming the river. Even the strident voice of the cricket, so discordant to some, fills his mind 'with a train of summer ideas, of everything that is rural, verdurous and joyful'. There is a continuity in his happiness; the same thoughts recur on the same occasions. 'I made the same remark in former years as I came the same way annually.' Year after year he was thinking of the swallows.

But the landscape in which this bird roams so freely has its hedges. They shut in, but they protect. There is what he calls, so aptly, Providence. Church spires, he remarks, 'are very necessary ingredients in the landscape.' . . . Besides Providence there are the castles and seats of the nobility. He respects them almost equally. The old families—the Howes, the Mordaunts—know their places and keep the poor in theirs. Gilbert White is far less tender with the poor—'We abound with poor,' he writes, as if the vermin were beneath his notice—than to the grasshopper whom he lifts out its hole so carefully and once inadvertently squeezed to death. Finally, shading the landscape with its august laurel, is literature—Latin literature, naturally. His mind is haunted by the classics. He sounds a Latin phrase now and then as if to tune his English. . . .

So we observe through our field-glasses this very fine specimen of the eighteenth-century clerical naturalist. . . . His observation of the insect in the grass is minute; but he also raises his eyes to the horizon and looks and listens. In that moment of abstraction he hears sounds that make him uneasy in the early morning; he escapes from Selborne, from his own age, and comes winging his way to us in the dusk along the hedgerows.

A clerical owl? A parson with the wings of a bird? A hybrid? But his own description fits him best. 'The kestrel or wind-hover,' he says, 'has a peculiar mode of hanging in the air in one place, his wings all the time being briskly agitated.'

> Novelist and essayist Virginia Woolf, in 'White's Selborne', *New Statesman and Nation* (30 September 1939), 460–1.

The slow approach of Autumn, avers our favourite Nature boy, explains that soft melancholy note in the song of the willow-wren. And what about the slow approach of the Nature boys?

Gilbert White of Selborne (we've often thought, reading between the lines) sometimes guessed he was not precisely *persona gratissima* with the birdies, and even realised, possibly, that one or two sad bird-songs were a personal crack at him, on behalf of the Ecclesiastical Commissioners. E.g., the song of the Lesser Spotted Guffin:

> Treading light on
> Buckled toe,
> Parsons come and
> Parsons go,
> Trapesing round the
> Blinking country,
> What's the point of
> Queen Anne's Bounty?
> Matings, calls, and
> Hibernations—
> What about them
> Congregations?
> Rev. G. White, this
> Goes for you,
> Ain't you got no
> Work to do?

> Writer and humorist (Dominic) Bevan Wyndham Lewis, in his weekly column 'Standing By . . .', *The Tatler* (20 August 1947), 242. (Queen Anne's Bounty was a fund set up in 1704 to augment the income of the poorer Anglican clergy.)

Anyone who lives in the country can experience at times, or seem to experience, an unmediated nature: in a direct and physical awareness of trees, birds, the moving shapes of land. What is new in Gilbert White, or at least feels new in its sustained intensity, is a development from this; a single and dedicated observation, as if the only relationships of country living were to its physical facts. It is a new kind of record, not only of the facts, but of a way of looking at the facts: a way of looking that will come to be called scientific: . . . it is not what can easily be confused with it from many earlier and some later observations, the working of particular social or personal experience into the intricacies of things seen. . . . Gilbert White was watching the turn of the year and the myriad physical lives inside it: nature in a sense that could now be separated from man.

Literary critic Raymond Williams, in *The Country and the City* (London, 1973), 118–19.

POSTHUMOUS LETTER TO GILBERT WHITE

It's rather sad we can only meet people
whose dates overlap with ours, a real shame that
you and Thoreau (we know that he read you)
never shook hands. He was, we hear, a rabid

Anti-Clerical and quick-tempered, you the
quietest of curates, yet I think he might well have
found in you the Ideal Friend he wrote of
with such gusto, but never ran into.

Stationaries, both of you, but keen walkers,
chaste by nature and, it would seem, immune to
the beck of worldly power, kin spirits,
who found all creatures amusive, even

the tortoise in spite of its joyless stupors,
aspected the vagrant moods of the Weather,
from the modest conduct of frogs to
the coarse belch of thunder or the rainbow's

federal arch, what fun you'd have had surveying
two rival landscapes and their migrants, noting
the pitches owls hoot on, comparing
the echo-response of dactyls and spondees.

Selfishly, I, too, would have plumbed to know you:
I could have learned so much. I'm apt to fancy
myself as a lover of Nature,
but have no right to, really. How many

birds and plants can I spot? At most two dozen.
You might, though, have found such an ignoramus
a pesky bore. Time spared you that: I
have, though, thank God, the right to re-read you.

Anglo-American poet W. H. Auden, 'Posthumous Letter to Gilbert White',
in *Thank You, Fog: Last Poems by W. H. Auden* (London, 1974), 32–3.

EXPLANATORY NOTES

Place of publication is London, unless otherwise indicated.

ABBREVIATIONS

GK   'Garden Kalendar', *The Journals of Gilbert White, 1751–93*, ed.
              Francesca Greenoak, 3 vols. (1986–9), vol. i.
LL   Rashleigh Holt-White, *The Life and Letters of Gilbert White of
              Selborne*, 2 vols. (1901).
NJ   'Naturalist's Journal', *The Journals of Gilbert White, 1751–93*, ed.
              Francesca Greenoak, 3 vols. (1986–9).
Phil. Trans.   *Philosophical Transactions of the Royal Society of London.*

TITLE PAGE

1 [*epigraphs*]: the main title page for the 1789 edition of *The Natural History
    and Antiquities of Selborne* carried two classical quotations. The first is
    from Horace's *Odes* (Book 4, Ode 2): 'I, like a bee on Mount Matinus,
    industriously engage in a favourite activity.' The second is from
    Giovanni Antonio Scopoli's *Annus I historico-naturales* (Leipzig, 1769):
    'To attempt a description of everything there is in the world, whether
    of God's work or wrought by the forces of created nature, is a fruit-
    less endeavour far beyond the skills, or even the lifetime, of one man.
    For this reason the most useful task is the preparation of books on the
    fauna and flora of a particular region: it is monographs that are of the
    greatest value.'
      The separate title page for the natural history part of the 1789 edition
    carried two further classical quotations. The first is from Homer's *Odyssey*
    (Book 9): 'It is a rough country, but a good place for rearing young men.
    I for my part cannot envisage anything sweeter than one's own country.'
    The second is from Cicero's *Oratio Pro Plancio* (Book 9): 'In short the
    whole of that territory of ours is rough and mountainous, [but] depend-
    able, trustworthy, and a protector of her own people.'

ADVERTISEMENT

3 *stationary men*: those who resided in the same place all the year round; the
    gentry and aristocracy usually left their country residences to spend the
    winter in London or to travel.

county of Southampton: an administrative area that controlled the geo-
    graphical territory of Hampshire.

acknowledgments . . . due: Magdalen College, Oxford, was Lord of the
    Manor of Selborne and keeper of the documents concerning the
    parish; White was helped by Richard Chandler, fellow of Magdalen

College, and also by Ralph Churton of Brasenose College. The president of Magdalen College was George Horne.

LETTERS TO PENNANT

7 *Hanger*: a wood on the side of a steep hill or bank.

freestone: stone that can be cut in any direction and readily shaped with a chisel.

8 *covert*: a thicket, which gives shelter to wild animals or game.

Nore . . . German ocean: Tacitus referred to the North Sea as the *Oceanus Germanicus*; the Nore is a sandbank in the Thames estuary and was one of the bases of the North Sea fleet.

9 *Wolmer-forest*: now known as Woolmer Forest; White used both names but in the errata to the first edition of *The Natural History and Antiquities of Selborne* he instructed that all references be changed to Wolmer.

shakey: fissured.

lime and turnips: farming on poor land was made possible by introducing turnips into the crop-rotation system to provide winter fodder for livestock as well as to clear the ground of weeds; lime improves fertility by reducing the acidity of sandy soils.

wych hazel . . . 1703: on 7–8 December 1703, the most severe storm recorded in southern Britain caused between 8000 and 15,000 deaths from flooding, shipwrecks, and destruction of buildings. William Derham took barometric readings during the storm, and Daniel Defoe published an account of it in *The Storm* (1704). 'Wych hazel' was used as a synonym for the wych elm (*Ulmus glabra*).

vulgarly . . . Plestor: the term in common usage for the 'Pleystow', an area for recreation similar to a village green; see *Antiquities of Selborne*, plate 9 and letter 10.

the vicar: White's paternal grandfather, Gilbert White, was vicar of Selborne from 1681 to 1728.

10 *the bridge . . . repairs*: the Hampton Court Bridge, near the Toy Inn, was the largest Chinoiserie bridge ever built; it opened in 1753.

eyry: brood in a nest.

11 *cardo*: hinge of a bivalve.

museum . . . article: the museum in Leicester House, London, contained Sir Ashton Lever's private natural history collections, including over a thousand fossils; it opened in 1774.

This bivalve . . . Gorgonia: *Gorgonia* was the name given to the coral known as the sea fan; 'zoophyte' was the term used to denote natural objects that appeared to be partly animal and partly plant.

drawn and engraved: see plate 3 in *The Natural History of Selborne* (1789).

Cornua Ammonis: the common ammonite.

inclining . . . Hanger: this path, known as the Bostal, was designed by White and his brother Thomas in 1780 to provide a less steep climb through the hanging woods than the zigzag path that White and his brother John had constructed in 1753.

recent production: there had been suggestions in the eighteenth century that the earth was eternal or extremely ancient, but the orthodox religious view maintained that it was around 6,000 years old; White's 'recent' would be relative to this short period of time, not to the modern geological time scale.

nautili: species of cephalopod (molluscs).

12 *Portland*: limestone from the Isle of Portland, Dorset, widely used as a building material.

Dr Plot. Oxfordsh. p. 77: Robert Plot, *The Natural History of Oxfordshire*, 2nd edn. (Oxford, 1705).

Plot's Staff, p. 152: Robert Plot, *The Natural History of Stafford-shire* (Oxford, 1686).

13 *terrene*: of the earth.

furlongs: a furlong is one-eighth of a mile.

14 *filices*: ferns.

affect: inhabit.

tread the bounds: the method of establishing the extent of a parish for ecclesiastical administrative purposes by walking the boundaries.

intelligent gentleman: White's brother-in-law Thomas Barker of Lyndon Hall, Rutland.

15 *bark timber*: bark was peeled from timber for use in tanning to produce water-repellent leather.

Plot . . . counties: Robert Plot, *The Natural History of Stafford-shire* (Oxford, 1686), 216, claimed only that no one had ever heard of them.

16 *shooting flying*: grouse and partridge fly close to the ground, requiring low, angled shots, but pheasants fly high and require 'shooting flying'.

Hales's Haemastatics: Stephen Hales, *Haemastatics* (1733); it formed the second volume of Hales's *Statical Essays*, which went through several editions.

17 *link . . . wanting*: a reference to the great chain of being, the idea that all beings and matter are organized in a strict hierarchical manner according to degrees of perfection, from minerals up to God.

Waltham blacks: one of the most famous groups of Hampshire poachers who blacked their faces when on raids.

prickers: horsemen.

Mr Astley's riding-school: Philip Astley was renowned for the acrobatic equestrian shows that were performed at his riding school near Westminster Bridge, London, from 1773.

17 *devoted deer*: the deer selected to be prey.

18 *Waltham blacks . . . before*: the deliberate targeting of the property of Charles Trimnell, Bishop of Winchester, by the Waltham Blacks during the economic difficulties of the early 1720s led him to instigate the passing of the notorious Black Act of 1723, which introduced the death penalty for being found in a forest wearing a disguise, and for forty-nine other offences.

Hoadly: Benjamin Hoadly was Bishop of Winchester from 1734.

lurcher: greyhound–collie cross; a hunting dog whose stealth and speed were valued by poachers.

19 *Candlemas and Midsummer*: Candlemas (2 February) was taken to mark the midpoint of winter, as it falls approximately half way between the winter solstice (21 December) and the spring equinox (20 March); midsummer's day falls on 24 June.

brouze: young shoots and twigs that provided fodder for cattle.

gentleman . . . Andover: Henry White, one of White's brothers, was rector of Fyfield, near Andover.

20 *keepers . . . Barnabas*: possibly an ancient midsummer ritual, as St Barnabas' day falls on 11 June, which was the longest day of the year before the calendar reform of 1752.

carex cespitosa: in J. D. Hooker's *The Student's Flora* (1884), 454, *Carex coespitosa* (given as a synonym for *Carex stricta*) is described as growing in large dense tufts, in marshy places from the Clyde to Kent and Dorset.

purlieu: a tract of land on the border of a forest; specifically, one formerly included within the forest boundaries and still partly subject to the forest laws relating to the hunting or killing of game.

hassocks: this term for tufts of grass or rushes was possibly applied to the cushions for kneeling on in church because they were stuffed with these rushes; before cushions came into use, the tufts themselves were trimmed and laid on the church floor for use as kneelers.

21 *ostensible*: formally agreed.

hungry . . . naked sand: sterile and barren; not supplying food for fish.

nice: exact, accurate.

22 *decoy*: a pond or pool out of which run narrow arms or 'pipes', covered with netting or other contrivances, into which wild ducks or other fowl may be lured and so caught.

prince . . . mezzotinto: the mezzotint print-making method was invented by the German artist Ludwig von Siegen; Prince Rupert, however, introduced the art into England, and invented the mechanism by which the engraved copper plate was roughened to give tonality to the print.

23 *lop and top*: the smaller branches and any waste timber cut from trees; claimed by the poor according to ancient custom.

run: bleed sap.

24 *fauna . . . October*: from 'Flora Selborniensis', *The Journals of Gilbert White, 1751–93*, ed. Francesca Greenoak, 3 vols. (1986–9), i, 219–20.

25 *alauda trivialis . . . motacilla trochilus*: the grasshopper-lark and the willow-wren (see letter 16 to Pennant).

in your Zoology . . . fly-catcher: White refers to the first edition of Thomas Pennant, *British Zoology* (1766), 99.

Derham . . . three: see *Philosophical Letters between the Late Learned Mr. Ray and Several of his Ingenious Correspondents*, ed. William Derham (1718), 108 n.

Stillingfleet . . . not: see Benjamin Stillingfleet, 'The Calendar of Flora', *Miscellaneous Tracts in Natural History*, 2nd edn. (1762), 298 n.

town: the meeting with Pennant occurred during White's visit to London ('town') from 18 April to 12 June 1767.

non-descript: not yet described and named in scientific literature.

26 *falco . . . Wales*: the falcon specimen was sent to Pennant's home in Wales.

Spring-Gardens: the Great Room in Spring Gardens, near St James's Park, London, was a fashionable venue for concerts and exhibitions; in 1764 the young Mozart performed before the King here.

cere: the area on the upper part of the beak that contains the bird's nostrils.

outlet: area such as a field or orchard attached to a house.

27 *squab-young*: newly hatched birds before they acquire feathers.

28 *chimnies . . . bacon*: bacon was hung to dry and to be cured in the smoke of wood fires in the wide chimneys of country cottages.

29 *mice . . . non-descript*: see note to p. 25; these were specimens of the harvest mouse, a species that White was the first to identify and describe.

30 *remiges*: quill feathers in a bird's wing.

Ray's . . . 1685: see *Philosophical Letters between the Late Learned Mr. Ray and Several of his Ingenious Correspondents*, ed. William Derham (1718), 183.

congeners: members of the same kind or class, related according to scientific classification.

Sunbury: the Surrey home of White's friend John Mulso, who was vicar there from 1746 to 1760.

osier-beds . . . aits: places where willows grow on small islands in rivers.

Now . . . September: Linnaeus believed that swallows hibernated at the bottoms of lakes during the winter; Alexander Malachias Berger, one of Linnaeus's students, stated that the swallow 'goes under water' in September (see Benjamin Stillingfleet, 'The Calendar of Flora, Swedish and English', *Miscellaneous Tracts in Natural History*, 2nd edn. (1762), 284).

30 *Borough*: the borough of Southwark, on the south bank of the Thames outside the City of London; White's brother Thomas was a wholesale merchant there.

31 *Now . . . equator?*: Michel Adanson had stated that four European swallows lodged on the ship in which he travelled from Gorée to Senegal in October 1749, and that swallows were seen only at that time of year in Senegal (Benjamin Stillingfleet, *Miscellaneous Tracts*, 2nd edn. (1762), 284 n.; the information was taken from Adanson's *Voyage au Sénégal* in the English translation (London, 1759), 121). Pennant claimed that Adanson's observations 'proved beyond contradiction' that some swallows migrated (see *British Zoology*, 2nd edn., ii (1768), 249).

32 *Stillingfleet . . . them*: see Alexander Malachias Berger's 'Calendarium florae' in Benjamin Stillingfleet, 'The Calendar of Flora, Swedish and English', *Miscellaneous Tracts in Natural History*, 2nd edn. (1762), 265 n.

(and . . . parts): White made an annual visit to Ringmer, near Lewes, to visit his aunt, Rebecca Snooke.

friend . . . war: James Gibson, clergyman at Bishop's Waltham, had served as a naval chaplain in North America and had seen service at Louisberg and Quebec during the Seven Years War.

33 *Hasselquist . . . Levant*: Frederik Hasselquist, a pupil of Linnaeus, was sent by him to study natural history in the Middle East.

Andalusia . . . there: White obtained his information about the Spanish climate from his brother John, who was chaplain to the garrison stationed at the strategic British naval base of Gibraltar. John White had gone to Gibraltar in 1756 as a form of exile, to escape being sued for debt in the period after he had been expelled from Corpus Christi College, Oxford, for aiding a fellow student to marry an innkeeper's daughter.

I have . . . Sunbury: John Mulso moved from Sunbury, Surrey, in 1760 to become vicar of Thornhill, Yorkshire.

avoirdupois: avoirdupois weight was the standard system of weights used in Britain.

lumping weight: good or full weight.

34 *There . . . 1739–40*: the winter of 1739–40 was one of the most severe remembered; the River Thames was frozen from Christmas day until 17 February.

vents . . . nose: these are scent glands; they are not used for breathing.

35 *mustelinum . . . cane*: the 'cane' is probably a female weasel, which is much smaller than the male; weasels belong to the family Mustelidae.

emberiza nivalis . . . Brit. Zool.: a reference to the first edition of Pennant's *British Zoology* (1766), 112–13, in which he stated that in England Greater Bramblings were 'mistaken for white larks'. The *Emberiza nivalis* was also known as the 'snowflake' or the 'snow lark-bunting'.

36 *salicaria . . . soft-billed*: an insect-eating warbler.

I shot . . . aim: White had been a keen sportsman in his youth, frequently going shooting.

campaign: open country.

37 *fallows*: ploughed fields left uncultivated for a season.

pointers: dogs that point out the location of game to sportsmen.

drams: a dram was a unit of weight equivalent to one-sixteenth of an ounce in the avoirdupois system, and one-eighth of an apothecaries' ounce.

38 *haunted*: frequented.

Mr Johnson: Ralph Johnson of Brignall, Durham.

LINNAEI NOMINA: Linnaean names. White preferred to use John Ray's names for birds (see letters 1 and 2 to Barrington, pp. 95–101) as did Pennant (see *British Zoology*, 2nd edn., i (1768), p. xiii).

39 *ventriloquous*: from the stomach.

you pursue . . . fishes: Pennant was completing the third volume of his *British Zoology*, 2nd edn. (1769), on reptiles and fish.

40 *cryptogamia . . . plants*: 'the sexual system' refers to Linnaeus's classification, which was based on the male and female parts of flowers; 'cryptogamia' ('hidden sex') was the name given to non-flowering plants.

a quack . . . oil: charlatans would eat or pretend to eat toads (held to be poisonous) in order to exhibit their skill in expelling the poison and thereby advertise their supposed remedies.

kenning him: catching sight of him.

41 *Swammerdam . . . female*: Jan Swammerdam described in *The Book of Nature* (1758), 111 (translation of *Bybel der nature*, 2 vols. (Leiden, 1737–8)), how the male frog attached himself firmly to the back of the female by throwing his forelegs across her breast and interlacing his toes, 'in the same manner as people do their fingers at prayers'. Swammerdam conjectured that the male exerted pressure with his forelegs to promote the emission of eggs by the female, and that the egg mass was then fertilized by the discharge of sperm from the anus of the male.

oeconomy of Providence: 'the all-wise disposition of the Creator in relation to natural things, by which they are fitted to produce general ends and reciprocal uses' (Isaac J. Biberg, 'The Oeconomy of Nature', in Benjamin Stillingfleet, *Miscellaneous Tracts in Natural History*, 2nd edn. (1762), 39); the divine management of the world.

Merret . . . rana arborea: Christopher Merret's alphabetical catalogue of the British flora, fauna and fossils, *Pinax rerum naturalium Britannicarum* (1666), contained many inaccuracies. The tree frog (*rana arborea*) is not native to Britain.

mud inguana: an error for 'mud iguana', but White copied Ellis who wrote 'Mud-Inguana'; see *Phil. Trans.*, 56 (1766), 189.

41 *English lacerta . . . newt*: newts (salamanders) were considered to be reptiles and were classified with lizards in the genus *Lacerta*; it was not until the nineteenth century that they were redefined as amphibians.

sallad-oil . . . viper: Pennant included this information in his *British Zoology*, 2nd edn., iii (1769), 39. The remedy had been mentioned in *Phil. Trans.*, 39 (1735–6), 312.

42 *snakes*: White here refers to grass or common snakes, not vipers.

visit . . . house: White stayed with his brother Henry in Fyfield from 2 to 16 July 1768.

43 *Fleet-street . . . Mazel*: White was sending the fish specimens to his brother Benjamin, Pennant's publisher, whose business was in Fleet Street. Peter Mazell was Pennant's engraver; most of the illustrations of fish in the third volume of the second edition of Pennant's *British Zoology* were drawn by G. Wilkinson.

not forget . . . toads: Pennant was collecting information about this cure, which he discussed in an appendix to *British Zoology*, iii (1769), 332–40; one of the published sources on which he drew concerned a patient in Hungerford (ibid., 333).

44 *nostrum . . . emolument*: secret remedy for his own profit.

exclude: discharge.

three . . . notes: White was the first to differentiate between these three species; his description of their songs allows for easy identification in the field (see also letter 1 to Barrington, p. 95):

> Smallest willow-wren (chiffchaff, *Phylloscopus colybita*): chirps
> Middle willow-wren (willow warbler, *P. trochilus*): sweet plaintive note
> Largest willow-wren (wood warbler, *P. sibilatrix*): sibilous grasshopper-like noise.

peremptorily: with strong or positive belief or assurance, emphatically.

45 *butcher bird*: so called because it impales the bodies of insects and small vertebrates on thorns to make a larder; it is also known as the shrike.

gentleman from London: identified in the *Selborne* MS as 'my Brother from Fleetstreet' (Paul Foster, ed., *The Natural History of Selborne*, Oxford World's Classics (Oxford, 1993), 270, note to p. 52); this was Benjamin White, who evidently visited Selborne in autumn 1768.

46 *Lady-day*: the Feast of the Annunciation of the Blessed Virgin, 25 March.

large black warty lizard: the great crested newt (see note to p. 41).

buck's head . . . creation: see letter 14 to Pennant (p. 34); 'Mr —' was probably the surgeon and collector of anatomical specimens John Hunter, who had been elected Fellow of the Royal Society in the same year as Pennant.

gentleman in Sussex: John Woods, father-in-law of White's sister Rebecca, who lived in the village of Chilgrove, near Chichester.

47 *friend near Chichester*: see preceding note.

Naturalist's Journal: a notebook arranged into blank compartments for recording natural phenomena over one year, devised by Daines Barrington and published for the first time in 1767 by Benjamin White.

a bird . . . you: Pennant lived in Flintshire, Wales, far outside the area of the stone-curlew's breeding grounds.

outlet: see note to p. 26.

Stonehenge . . . impost stones: the ancient monument of Stonehenge in Wiltshire consists of a stone circle, in which pairs of massive upright stones support horizontal lintels ('impost stones'). White had visited Stonehenge in 1755.

bottom: low ground.

48 *You judge . . . toads*: see letter 18 to Pennant (p. 42).

icthyology: (White misspells 'ichthyology') the study of fish.

livings: benefices; in the Church of England, a parish or group of parishes under a single stipendiary minister.

49 *Northamptonshire . . . view*: White travelled in this flat part of the country in 1746, when he visited the Isle of Ely and Lincolnshire.

prospects: extensive or commanding views, vistas.

hams: plots of pasture ground or meadowland.

error . . . herons: Pennant's error had been to describe crested and uncrested herons as two species in *British Zoology*, 2nd edn., ii (1768); it was only on visiting the heronry at Cressi Hall that he realized that they were the male and female forms of one species (see Pennant, *A Tour in Scotland, 1769* (Chester, 1771), 14).

whose . . . Cressi-hall is: it was the seat of the Heron family; the heronry was probably established to celebrate the family name.

51 *draughtsman . . . Zoology*: Peter Paillou (*c.*1720–1790) drew the female goatsucker in the first edition of Pennant's work (1766), plate R1; the male is depicted separately (plate R), but the artist's name does not appear on this plate. The name 'goatsucker' was given to what is now known as the nightjar in the belief that it sucked the udders of goats.

Portsmouth evening gun: a gun was fired at Portsmouth, the main dockyard of the Royal Navy, every noon and evening; Portsmouth is approximately twenty miles from Selborne as the crow flies.

hermitage . . . building!: the churn owl was an alternative name for the goat-sucker (nightjar); the hermitage, a wooden thatched structure on Selborne Hanger, was a mock hermit's cell that was used as a summerhouse.

non-descript: not yet named or described in print.

Indian-grass . . . silkworm: Indian grass was an undefined fine and strong thread sold to anglers (see *The Gentleman Angler* (1726), 6); it was thought by Pennant to be the tendrils of the 'purse' or egg case of the greater dog fish (see *British Zoology*, 2nd edn., iii (1769), 89).

51 *vignette . . . book*: the illustration of the hermitage appeared on the title page of the 'Natural History' section of *The Natural History and Antiquities of Selborne* (1789).

52 *Guernsey . . . Oxford*: links were forged between Guernsey and Pembroke College, Oxford, in the seventeenth century, when, in order to bring Presbyterian Guernsey into conformity with the Church of England, Pembroke established fellowships and scholarships for those on the island wishing to train as clergymen. It is unclear why the bright green Guernsey lizard (*Lacerta viridis*) was introduced into the college garden. Pennant incorporated the information supplied by White here and in letter 22 (see p. 48) into *British Zoology*, 2nd edn., iii (1769), 14, and suggested that *Lacerta viridis* might be related to the sand lizard *L. agilis*, found in Britain.

tree . . . nests: it was an oak tree (see Pennant's *A Tour in Scotland, 1769* (Chester, 1771), 14).

caprimulgus: the goatsucker (see letter 22 to Pennant, p. 48, and notes to p. 51 above).

If . . . straggler: this paragraph was paraphrased by Pennant and published in *British Zoology*, 2nd edn., iv (1770), 12–13; White had witnessed the migration on 29 September (Michaelmas day) 1768, when travelling to Goodwood, near Chichester, to see the Duke of Richmond's moose (see letter 28 to Pennant, p. 65).

53 *Merton-wall*: the wall between Merton College, Oxford, and Christ Church Meadows; Pennant knew Oxford, having, like White, been a student there.

scarabaeus fullo . . . seacoast: *scarabaeus fullo* (now *Melolontha fullo*) is a cockchafer beetle; following his first meeting with Joseph Banks in May 1767, White had met him for a second time in London in May 1768.

54 *dressed*: prepared for cooking. As a young man, White had had a taste for exotic food (see Richard Mabey, *Gilbert White* (1986), 37).

These birds . . . counties: Pennant included White's observations in *British Zoology*, 2nd edn., iv (1770), 56–7, but still maintained that ring ouzels resided in the mountains of Wales and Scotland for the whole year.

Revesby, in Lincolnshire: Revesby Abbey and estate was Joseph Banks's country home.

My bird . . . locustella: White was the first to distinguish between these birds and to provide a description of the sedge warbler.

For . . . 1769: letter 25 to Pennant (see p. 57).

55 *The late . . . Atlantic*: in the letter sent to Pennant (as opposed to the version published in the *Natural History*), White mentioned Alexander Catcott, whose *Treatise on the Deluge*, 2nd edn. (1768), contained a discussion of 'The Time, when, and the Manner how America was first peopled'.

Amusive: engaging the attention, affording pleasing entertainment.

aurelia: chrysalis of an insect.

56 *Hero and Leander*: a story of ill-fated lovers in Greek mythology, told by both Ovid and Virgil, in which Hero lights a lamp in her window every night for Leander to use as a guide to swim across the Hellespont.

57 *crabbed*: difficult to understand.

58 *fen . . . Revesby*: Revesby, Joseph Banks's estate, is in the fens, a wetland area in the east of England.

letter . . . 1769: letter 24 to Pennant (see p. 53).

squnck, or stonck: skunk; a North American animal of the weasel kind. John Ray briefly mentioned the 'squnck' in the Preface (p. [vi]), to his *Synopsis methodica animalium quadrupedum et serpentini generis, vulgarium notas characteristicas, rariorum descriptiones integras exhibens* (1693).

A gentleman . . . drawing: White's brother John, stationed in Gibraltar, had sent him a specimen of the woodchat shrike, a rare vagrant to Britain; in *British Zoology*, 2nd edn., i (1768), 164, Pennant stated that since he had not seen this bird he was obliged to use a drawing made by George Edwards, kept in the Sloanian Museum, the core collection of the British Museum.

59 *baiting place*: a place where a halt is made for refreshment on a journey.

snow-fleck: White had asked about the *Emberiza nivalis* (greater brambling or snowflake) in letter 15 to Pennant (see p. 35).

61 *work*: White refers to the fourth volume of the second edition of Pennant's *British Zoology* (1770), which aimed to 'supply the deficiency of plates in the three first volumes'.

article of: so far as concerns, with regard to. White welcomed Pennant's discovery of the water shrew because, like the white hare, it was a species of quadruped new to Britain. It had been discovered by the French naturalist Louis-Jean Marie Daubenton (see Buffon, *Histoire naturelle*, viii (Paris, 1760), 64). In the eighteenth century, shrews were often called mice because of their size and appearance, but they are not rodents, as mice are.

hybernaculum: the winter quarters of a hibernating animal.

gallon: a unit of volume, equivalent to about 4.5 litres, that was used as a dry measure for bulk commodities such as grain.

63 *Andalusia . . . August*: White obtained this information from his brother John, who was chaplain in Gibraltar.

The great . . . England: White was the first to describe the noctule bat in England (see letter 36 to Pennant, p. 76); the French naturalist Georges-Louis Leclerc, Comte de Buffon had described it in *Histoire naturelle*, viii (Paris, 1760), 128–9.

phalaenae: *Phalaena* was a (now obsolete) Linnaean genus of moths.

defect: deficiency.

64 *hybernaculum . . . winter*: the nests in which hedgehogs hibernate are usually hidden under hedges or in the roots of trees.

as may . . . fauna suecica: Carl von Linné [Linnaeus], *Fauna Svecica*
(Stockholm, 1761), 78.

larkers . . . bat-fowlers: larkers caught larks by dragging nets over the
ground; bat fowlers caught roosting birds at night by using bright lights to
encourage the birds to fly out of hedges into nets, where they could be
struck down using 'bats' or staves.

somewhat: something.

65 *moose . . . Richmond*: this was the second young female moose owned by the
Duke (the first had died in 1767); by 1770 he also owned a yearling bull
moose, the subject of the famous painting by George Stubbs. It was hoped
that the study of the moose would contribute to the debate about whether
there was a living counterpart of the giant fossilized Irish elk, whose enormous horns and bones were occasionally dug up from peat bogs. Some
naturalists thought it to be an extinct species, but others, for whom total
extinction was impossible in a providential world in which all creatures
formed essential links in the great chain of being, believed that a living
form must continue to exist somewhere.

wither: the withers are the highest part of the back, lying between the
shoulder-blades.

67 *and whether . . . creature*: the following year, Pennant stated in print that
the similarity of the horns of the North American moose to those of the
European elk proved that they were the same species, the larger size of
the moose being due only to availability in the American forests of more
'luxuriant food' (see *Synopsis of Quadrupeds* (Chester, 1771), 41–2). He did
not believe that either was related to the fossil Irish elk because of differences in their horns, but suggested that a living counterpart of the giant
fossil might be the enormous moose-like 'waskassou', known only from
reports by the Native Americans in Hudson's Bay (ibid., 44–5).

69 *setting*: hunting with a dog trained to stand still and indicate the position
of game by pointing with its muzzle.

male otter . . . me: this information was not in the original 1770 letter to
Pennant; its inclusion in the published version shows how White brought
together materials on the same topic when working on the *Natural History*.
The otter was killed on 3 October 1784, and was the first found in the
parish (NJ).

Scopoli's . . . see it: the first volume of Scopoli's *Anni historico-naturales*,
a study of the natural history of the province of Carniola, was published in
1769; for his *Entomologia*, see note to p. 110.

the chaplain: James Gibson, naval chaplain in North America (see letter 13
to Pennant, p. 31).

70 *An . . . gentleman*: Sampson Newbery of Exeter College, Oxford, whose
father lived in Devonshire.

71 *conversation*: the action of consorting or having dealings with others, living together.

advances . . . fact: the woodcock does not carry its young in its beak, but it can carry one chick at a time between its thighs.

72 *forestalled . . . non-descript*: the discoverer of a new species is the person who first describes it in print; White had received a specimen of a winter swallow from Gibraltar, sent by John White, and, supposing it to be a new species, had named it *Hirundo hybernia*; he subsequently discovered that this bird was identical with Giovanni Antonio Scopoli's *Hirundo rupestris* (Mediterranean crag martin) (see *LL*, i, 183, 194).

Barbary: the term used by Europeans for the coastal regions of North Africa (Morocco, Algeria, Tunisia, and Libya).

characters: the distinctive features by which a natural object is classified.

manners: habitual mode of behaviour.

73 *Sussex friend*: John Woods (see letter 21 to Pennant and note to p. 46).

74 *my relation*: White's brother, John.

Warreners: keepers of warrens, where a breeding stock of rabbits was kept.

75 *coleoptera*: beetles.

Oestrus: gadfly: specifically a reference to the gadfly sent by Zeus's wife Hera to torment Io, a nymph seduced by Zeus, and changed by him into a heifer.

Geoffroy, t. 17, f. 4.: the reference is to Étienne-Louis Geoffroy, *Histoire abrégée des insectes qui se trouvent aux environs de Paris*, 2 vols. (Paris, 1762), plate 17, figure 4. White considered the plates in Geoffroy's work the best he had ever seen (see Paul Foster, 'The Gibraltar Correspondence of Gilbert White', *Notes & Queries*, 32 (1985), 323).

76 *uropygium*: the rump in birds.

calculus aegogropila: a misspelling of *Calculus aegagropila*, a hairball that becomes hard and glossy owing to continual friction within the stomach. White noted in NJ, 1 March 1770, that the *Calculus aegagropila* was described in Linnaeus, *Systema naturae*, iv (1770), 176, item 5.

77 *drachm*: dram; one-sixteenth of an ounce (see note to p. 37).

In the extent . . . smell: White was the first person in England to describe the noctule bat.

78 *fern-owl*: another name for the goatsucker or churn-owl (nightjar).

fern-chafers: the fern-chafer is a species of beetle.

evolutions: aerial wheeling and diving.

79 *lake . . . hybernaculum*: the Swedish naturalist Linnaeus and his followers believed that swallows hibernated in shelters at the bottoms of lakes.

cantoned: divided into small groups.

80 *new . . . British Zoology*: by 1773, Pennant was working on the fourth edition of *British Zoology*, 4 vols. (1776–7); it included many of White's observations.

80 *osprey*: White separately footnoted most of the birds mentioned in this letter and gave the following page references to Pennant's *British Zoology*, 4 vols. (1768–70), i: osprey, 128; butcher-bird, 161; crows, 167; Cornish choughs, 198; wild-pigeon, 216; missel-bird, 224; ring-ousels, 229; ii: tit-larks, 237; swallows, 242; house-martins, 244; swift, 245; whin-chats and stone-chatterers, 270, 271; wheat-ears, 269; bulfinches, 300; chaffinches, 306; cock-snipes, 358; lapwings, 360; little auk, 409; teals, 475; iv: swift, 15; sedge-bird, 16.

81 *Adanson's . . . species?*: White had already expressed doubts about Adanson's testimony in letter 12 to Pennant (see p. 29). Adanson had supported his argument that the swallows in Senegal were European migrants by pointing out that they did not build nests there; White's evidence to the contrary was obtained from Charles O'Hara, commandant of the Africa Corps in Senegal in 1766.

ventriloquous: from the stomach.

page . . . dew: White quotes from Pennant's *British Zoology*, 2nd edn., iv (1770), 15.

82 *pulli*: unfledged young birds.

nidification: nest building.

individuals . . . chimnies: swallows are prone to nesting in wide chimneys.

The threads . . . mysterious: the 'threads' are now known to be parasitic worms. The reproduction of European eels is still not fully understood, but it is thought that, when about 15 years old, they undertake a remarkable migration to the Sargasso Sea (in the Atlantic near the Bahamas) in order to mate; the adults then die and the transparent eel larvae take about three years to float on oceanic currents back to Europe where these 'glass' eels enter the river systems.

83 *jaded*: worn out, exhausted.

flirt: to move with a jerk or spring; a darting motion.

85 *affects neighbourhoods*: inhabits places of human habitation.

sweetbriar: a species of rose.

retires: withdraws or migrates.

LETTER 41: this entire letter was originally part of a letter sent to Barrington, dated 15 January 1770; White changed the addressee and included it here because the content followed on from that of letters 39 and 40 to Pennant.

imbecility: weakness.

86 *aurelia*: chrysalis.

Phryganeae: caddisflies; the chrysalises are found on aquatic vegetation or stones in shallow water.

sinks: open drains, pits for waste water.

lepidoptera ordo: the order of lepidoptera (butterflies and moths).

87 *warrens*: areas of land enclosed and preserved for breeding game.

faunist: one who studies animals (fauna) of a particular country or district.

improvements . . . us: the Anglo–Irish elite who governed Ireland in the eighteenth century made a profound impact on the Irish economy through their agricultural improving activities; their political assertiveness and their attitudes towards their Catholic estate tenants, however, led to distrust by the Irish population of the motivation behind this improvement. Premiums were paid to those who reclaimed wasteland or to those who grew crops favoured by their landlords.

88 *The military . . . omitted*: George Wade, commander of the army in Scotland from 1724 to 1740, had constructed over 250 miles of military roads to open up the Highlands, as well as building Fort Augustus and Fort George in order to control the Jacobites; Fort William had been built earlier. Herman Moll's map of Scotland was first published in 1714.

zigzag . . . Coryarich: Pennant had described the zigzag road cut into the steep side of the high mountain Coryarich, near Fort Augustus, in his *A Tour in Scotland, 1769* (Chester, 1771), 222.

policy: enclosed grounds or park surrounding a large country house.

cere: see note to p. 26.

90 *remiges*: see note to p. 30.

Sir Roger . . . Caernarvonshire: Roger Mostyn lived in Mostyn Hall, Flintshire, but the family owned other properties; the dovecote was at Gloddaeth Hall, near the limestone headland of Great Orme in Caernarvonshire.

fastnesses: places of safety and security from invasion.

91 *Their . . . sound*: White misquotes Milton's line, 'Their rising all at once was as the sound'.

mast: fruit of woodland trees.

LETTERS TO BARRINGTON

95 *last month*: White was in London from 15 April to 13 May 1769.

96 *amusing*: engaging the attention in a pleasing way.

as . . . numerically: that is, according to the numbers in White's list immediately above.

97 *old Michaelmas*: 10 October; Michaelmas is celebrated on 29 September, but after the change to the Gregorian calendar in 1752, which moved on the dates by eleven days, 10 October (29 September in the Julian calendar) became known as 'old Michaelmas day'.

101 *Ascension . . . boobies*: boobies are seabirds of the Sulidae family that are especially prolific on Ascension Island; their name was given them by seamen, who found that the birds were so tame they allowed themselves to

be taken by hand like boobies (fools). See Hans Sloane, *A Voyage to the Islands* (1707), i, 31.

102 *And . . . Shakespeare*: White misquotes 'And turn his merry note | Unto the sweet bird's throat', from *As You Like It*, Act 2, scene 5.

polyglot: a user of many languages; in this context, a bird that imitates the songs of other birds.

103 *Cherwell*: one of the rivers that flow through Oxford; White was in Oxford from 11 to 13 July 1769.

105 *What . . . ostrich*: a reference to Job 39: 13–16, in which the lack of maternal affection in the ostrich is attributed to the fact that she does not sit on her eggs to incubate them.

Zoology . . . 16: White refers to Thomas Pennant's *British Zoology*, 2nd edn., ii (1768), 320, and iv (1770), 16, where 'lesser reed sparrow' is given as a synonym for 'sedge-bird'.

106 *insensible perspiration . . . summer*: insensible: imperceptible; eighteenth-century medics considered that preventing weight loss through insensible perspiration was essential for good health.

supposititious: put by devious means in the place of another.

Your . . . suspicion: Daines Barrington published the results of dissections of the larynxes of birds (undertaken by the anatomist John Hunter), in his article 'Experiments and Observations on the Singing of Birds', *Phil. Trans.*, 63 (1773), 262, 278.

autopsia: the act of seeing with one's own eyes, personal observation.

107 *Investigations . . . occurred!*: Barrington referred to Virgil's descriptions in the *Georgics* of frozen rivers in Calabria in the southern part of Italy, in 'An Investigation into the Difference between the Present Temperature of the Air in Italy and Some Other Countries, and what it was Seventeen Centuries ago', *Phil. Trans.*, 58 (1768), 65.

108 *quicksilver mines*: Carniola is one of the main areas for mining highly toxic cinnabar, which is then heated to extract mercury (quicksilver).

109 *Kuckalm . . . Jamaica*: Tesser Samuel Kuckahn discussed the method he had developed for preserving bird specimens in hot climates in *Phil. Trans.*, 60 (1770), 302–20; he also presented specimens to the Royal Society.

110 *Scopoli . . . Entomology*: Giovanni Antonio Scopoli had published on the insects of Carniola, *Entomologia Carniolica* (Vienna, 1760), before publishing his series of *Anni historico-naturales* (Leipzig, 1769–72), which included descriptions of birds from various collections.

characters: the distinctive features by which a natural object is classified.

111 *This district*: White was on his annual visit to his aunt, Rebecca Snooke, who lived in Ringmer, on the South Downs.

suspicious . . . cantoned: suspicious: appears likely; cantoned: divided into small groups.

Brighthelmstone: Brighton.

113 *A land tortoise . . . visiting*: this was Timothy, Rebecca Snooke's tortoise.

intercourse: social communication between individuals.

commerce: sexual intercourse.

Fauna . . . 318: works by Linnaeus (see notes to pp. 64 and 76).

114 *wheat-sowing . . . oats*: that is, from September to March.

Sure . . . things: Barrington used White's comments about woodcocks as evidence against migration in his 'Essay on the Periodical Appearing and Disappearing of Certain Birds, at Different Times of the Year', *Phil. Trans.*, 62 (1772), 306–11.

faunists: those who study the animals (fauna) of a particular country or district.

116 *There fell . . . half*: this information was supplied by Thomas Barker, White's brother-in-law.

brother in Andalusia: John White.

117 *Old Belon . . . vultures*: see Pierre Belon, *L'Histoire de la nature des oyseaux* (Paris, 1555).

blood . . . food: the ancient Greek view that the body is made up of humours (bodily fluids), affected by physical qualities (hot/cold, wet/dry), was still prevalent in the eighteenth century; everything that was eaten was believed to have an inherent humoural temperature that needed to be balanced within the body.

German ocean: the North Sea (see note to p. 8).

118 *dreadful . . . 1708–9*: this was an extraordinarily cold winter across Europe; William Derham's record of temperatures was published in *Phil. Trans.*, 26 (1753), 454–78.

alauda pratensis: White's titlark (see letter 7 to Barrington, p. 109).

Tunstal . . . birds: Marmaduke Tunstall was a bird collector, who had a museum in Welbeck Street, London; he and Barrington saw White's specimens, sent from Gibraltar by John White and received by his brother Thomas White, a London merchant.

119 *My . . . visiting*: this was White's brother Henry, whom White visited from 5 to 22 February 1771.

The pipe . . . pitch: standard musical pitch was established only in the nineteenth century; pitch was lower and more variable in the eighteenth century, hence the need for White to describe the pipe he used, despite stipulating in the previous letter that it was set at 'concert-pitch'.

122 *approximation*: proximity, closeness.

daws: birds of the crow kind, but smaller, commonly called jackdaws.

us: White and his brother Thomas had both visited Ringmer on 1 November 1771.

123 *While . . . you*: see letter 7 to Barrington (p. 109), written from Ringmer.

124 *good old lady*: Rebecca Snooke, White's aunt.

   sublimes: elevates, heightens.

125 *Zoology . . . Ray*: see Pennant's *British Zoology*, 2nd edn., ii (1768), 264–5.

   jealousy: solicitude or anxiety for the preservation of something; vigilance in guarding a possession from loss or damage.

126 *hotbed . . . dung*: a hotbed is a bed of soil enclosed in frame, which is heated by fermenting manure and is used for raising or forcing plants; as a melon and cucumber grower, White needed hotbeds, and annually used up to fifty cart loads of dung for this purpose.

   preposterous: contrary to nature, perverse, monstrous.

   stroke: achievement, action towards an end.

127 *address*: skill, dexterity.

   remiges: see note to p. 30.

   congeries: a collection of things massed or heaped together.

128 *bushels*: a bushel was a unit of volume of eight gallons, used as a dry measure for agricultural commodities (see note to p. 61).

   It will . . . Transactions: White's four letters to Barrington were published in *Phil. Trans.*, 64 (1774), 196–201; 65 (1775), 258–76.

   outlets: the outlying land around a property.

   Ulloa's travels: see Antonio de Ulloa, *A Voyage to South America*, trans. John Adams, 2 vols. (1758), i, 174, who described how insects would fill the 'eyes, ears, and nostrils' of the inhabitants of Guayaquil, Ecuador.

129 *dipterous*: two-winged.

   subulated: slender and tapering.

   pupae: a pupa or chrysalis is the transitional stage between larva and adult in insects that undergo metamorphosis.

   Tom. iv, pl. II: White refers to René-Antoine Ferchault de Réaumur's *Mémoires pour servir à l'histoire des insectes*, 6 vols. (1734–42), iv, plate 11.

   monography: a monograph; a treatise on a single species, genus or larger group of animals or plants.

130 *tread*: to copulate.

132 *close air of London*: the air quality in London was notoriously poor owing to the number of domestic coal fires.

   Borough: the borough of Southwark (see note to p. 30).

   nestlings . . . Michaelmas: nestlings are young birds not old enough to leave the nest; 'unfledged' refers to newly hatched birds not yet covered in feathers. Michaelmas falls on 29 September.

133 *monography*: a reference to White's previous letter to Barrington on the house martin.

respectable . . . them: White's letter on the house martin was read to the Royal Society of London on 10 February 1774; Barrington had been elected a fellow of the Society in 1767.

majestic mountains: the Sussex downs are part of a chalk escarpment.

wild, or weald: the Weald is a clay area that is situated between the parallel chalk escarpments of the North and South Downs.

134 *plastic power*: an active life force delegated by God. This was one of the concepts that characterized the theology based on the constitution of the natural world (physico-theology), developed by John Ray.

flocks of Laban . . . other: see Genesis 31: 32–42.

135 *friend . . . Chichester*: probably John Woods (see note to p. 46).

137 *Shrove Tuesday . . . February*: the date of Shrove Tuesday is variable as it is tied to the date of Easter; White's boyhood dated from before the calendar reform in 1752, when the change from the Julian to the Gregorian calendar resulted in dates being moved forward by eleven days, and when 1 January instead of 25 March became the start of the year.

138 *address*: skill, dexterity.

139 *complacency*: satisfaction, pleasure, delight.

141 *Sir Ashton Lever's Musaeum*: in letter 3 to Pennant (see p. 11), White mentions visiting this museum to examine a fossil; it opened in London in 1774, but before that it was located in Lever's Lancashire home, Alkrington Hall. John White had been Lever's neighbour since leaving Gibraltar and becoming vicar of Blackburn, Lancashire, in 1772; it was John who told White about the curious nests (see *LL*, i, 238).

142 *Juturna . . . Aeneas*: Juturna, the warrior maiden of Virgil's *Aeneid*, who took the reins of her brother's chariot and guided it so that it flew like a swallow through a great house in order to escape Aeneas.

pamphlets . . . mislead: correspondence in newspapers claimed that farmers united to fix prices before the corn got to market; bread was the staple of labouring people in this period and the high price of corn prompted rioting in many areas.

143 *cunabula*: cradles.

terebrates: 'terebrate' means to form by boring.

144 *inartificially*: clumsily.

stools: stands.

species is cryptogame: it breeds in situations where it cannot be observed.

146 *Near Valencia . . . Montagna*: see John Ray, *The Ornithology of Frances Willughby* (1678), 213.

148 *mute*: defecate.

notorious: apparent, well known.

untiled . . . build: White recorded in NJ, 30 June 1774, that he 'procured a bricklayer to open the tiles in several places around the eaves of my Bro^{r's} brewhouse', so that he could examine the state of swifts' nests and the number of their young. White stayed with his brother Henry at Fyfield from 22 June to 7 July.

149 *still rain*: rain that falls in minute drops.

moult . . . breeding: swifts normally moult twice, once in the autumn and again in the spring.

151 *On the fifth . . . tedious!*: White added this paragraph when editing the letter for *Selborne*.

153 *eyes . . . Tobit*: White refers to Tobit 2: 10 in the Apocrypha: 'And I knew not that there were sparrows in the wall, and mine eyes being open, the sparrows muted warm dung into mine eyes.'

Devonshire . . . 1st: Sampson Newbery (of Exeter College, Oxford) sent the information from Devon (see *LL*, i, 253), and White's brother John was his informant in Lancashire.

Welch . . . llwyn: Pennant (who was Welsh) had mentioned this name in *British Zoology*, 2nd edn., i (1768), 224.

155 *hoodwinked*: blindfolded.

gentleman . . . veneration: this was White's father, John White.

meteor: an atmospheric phenomenon, such as mist or, as in this case, a gossamer shower.

156 *strange . . . formerly*: in NJ, 28 June 1775, White noted that Chaucer had considered gossamer as strange a phenomenon as thunder, tides, or mist. Gossamer is produced in the autumn when the webs spun by money spiders take flight as the sun warms the earth; they then fall to earth as the temperature drops in the evening. Before it was known that the floating gossamer was produced by spiders, it was believed by some to represent bits of the Virgin Mary's winding sheet that fell to earth during her assumption into heaven, or to be the substance of clouds or the finest down from geese (Peter Marren and Richard Mabey, *Bugs Britannica* (2010), 111).

apterous: without wings.

Dr Lister . . . fall: see the letter from Martin Lister to John Ray in *Philosophical Letters between the Late Learned Mr Ray and Several of his Ingenious Correspondents*, ed. William Derham (1718), 36–8. Lister held that the spiders carried on spinning in the air, which allowed them to change the dimensions of their webs, and thus to control their descent, like hot-air balloonists.

158 *lie-in*: give birth.

159 *Decayed*: fallen off, reduced in quality of health or prosperity.

Bell's . . . China: these were part of John Bell's *Travels . . . to Diverse Parts of Asia*, 2 vols. (Glasgow, 1763).

160 *medicated*: impregnated with a substance or medicine.

minuted: timed.

'*darkness visible*': an allusion to John Milton's *Paradise Lost*, I. 63.

farthing: a quarter of a penny.

162 *stools*: the stands of the beehives.

metheglin: a spiced alcoholic liquor made from fermented honey.

Thou . . . be: White adapted lines by John Philips in *Blenheim, A Poem* (1705), 14: 'Thou, His Associate Dear | Once in this World, nor now by Fate disjoin'd, | Had thy presiding Star propitious shone, | Shouldst CHURCHILL be!' Thomas Wildman had published *A Treatise on the Management of Bees* (1768). He was renowned for his performances with bees, achieved by his expertise in handling the queen bee in order to control the movement of swarms.

163 *cicatrices*: scars of healed wounds.

164 *manor, or hundred*: territorial units of jurisdiction.

The late . . . burnt it: White misquoted Jonathan Swift's line 'So the next parson stubbed and burnt it' from his *Baucis and Philemon: A Poem, on the Ever Lamented Loss of the Two Yew-Trees, in the Parish of Chilthorne* (1709).

way-warden: a person who supervises the highways of a parish or district.

alembics: an alembic was an apparatus consisting of two vessels, one containing a substance to be distilled and the other acting as a receiver in which the product condenses.

166 *Dr Hales . . . earth*: see Stephen Hales, *Vegetable Statics* (1727), 53; in later editions it formed the first volume of Hales's *Statical Essays*.

167 *Monsieur . . . belly*: see François-David Hérissant, 'Observation anatomique sur les organes de la digestion de l'oiseau appelé coucou', *Histoire de l'Académie Royale des Sciences* (1752), 417–23, and three plates.

168 *we . . . sun*: this occurred at Ringmer, where White and his nephew Samuel Barker were staying with Rebecca Snooke.

169 *Mr Lisle . . . him*: White refers to Edward Lisle's *Observations in Husbandry* (1757), 407.

170 *converse*: engage in sexual intercourse.

171 *leveret*: a young hare in its first year.

desiderium: a longing for a thing once possessed and now missed.

172 *Earth-worms . . . chasm*: all things, animate and inanimate, were thought to be part of a hierarchical sequence that constituted a great chain of being (see note to p. 17).

175 *The good . . . fish*: the notion that the experience of a pregnant woman, especially an obsessive longing, could be imprinted on her foetus was widely believed to account for birth defects; although this concept was increasingly discredited among eighteenth-century physicians, it continued to be widely held.

176 *marvellous account . . . May*: one such account appeared in David Hume's *The History of England*, 8 vols. (1767), ii, 381.

176 *sweetening . . . juices*: a reference to the need to balance the humours of the body in order to maintain health (see note to p. 117).

Green-stalls . . . fortunes: there was an increase in market gardens around London in the eighteenth century as exotic vegetables such as broccoli (introduced from Italy) and the quest for culinary novelty became increasingly fashionable among the well-to-do. Other vegetable novelties included forced produce; White himself managed to produce cucumbers (grown on hotbeds, see note to p. 126) as early as 5 April in 1763, noting that in London cucumbers cost two shillings apiece at this early date (in contrast to summer cucumbers that were two for 1½d.). The craze for unusual or out-of-season vegetables among the rich resulted in gluts of cheap vegetables for the poor in the summer months.

177 *hinds*: farm servants.

Potatoes . . . .reign: during the reign of George II, the suspicious associations of potatoes (introduced into Europe in the second half of the sixteenth century) included Catholicism and leprosy, as well as their botanical relation to deadly nightshade (see R. N. Salaman, *The History and Social Influence of the Potato* (Cambridge, 1985)).

Mr Ray . . . Europe: see John Ray, *Travels through the Low-countries, Germany, Italy and France*, 2nd edn., 2 vols. (1738), i, 348; Ray's journey lasted from 1663 to 1666.

sallets: salads.

178 *dactyls*: in poetry, a metrical foot consisting of a long syllable followed by two short ones.

179 *embarassed spondees*: embarrassed: confused; spondee: a metrical foot consisting of two long syllables.

centrum phonicum: in polysyllabic and articulate echoes, the place where the speaker stands (Charles Hutton, *A Mathematical and Philosophical Dictionary* (1796), i, 270).

180 *Anathoth*: the name of a biblical city, a place of echoes (see Isaiah 10: 30).

181 *quick-set*: hawthorn.

183 *monographies*: see letters 16, 18, 20, and 21 to Barrington (pp. 129, 137, 143, 146).

phytologist: botanist.

184 *spontaneous vegetation*: growing or produced naturally without cultivation or labour.

polar . . . species: see Georg Forster, *A Voyage Round the World*, i (1778), 511–15, for a description of cannibalism among the natives of New Zealand.

salutiferous: conducive to health.

swerd: the surface of the soil covered with grass.

'two . . . before': in Jonathan Swift, *Gulliver's Travels*, i (1726), 276, the King of Brobdingnag declares: 'that whosoever could make two Ears of

Corn, or two Blades of Grass to grow upon a Spot of Ground where only one grew before, would deserve better of Mankind, and do more essential Service to his Country, than the whole race of Politicians put together.'

185 *champaign*: level, open country.

filices: ferns.

186 *Hudsoni*: the presence of this latinized name after the Latin binomial for the yellow wort indicates that the plant was named and described by William Hudson, whose *Flora Anglica* (1762) was the first handbook on British wild plants to be based on the Linnaean system; few copies of this work were published, but White had purchased one in 1765 just before it sold out.

blowing: a state of blossoming; to be in bloom.

187 *corolla*: the whorl of coloured (i.e. not green) leaves (petals) that form the inner part of the flower.

Of all the propensities . . . delay: this section of the letter was included in the first edition of *Selborne* after the 'Antiquities', with the heading: 'Additions to Letter 41'.

188 *air*: the characteristic behaviour and motion of a bird that makes it instantly recognizable to the practised observer.

190 *remiges*: see note to p. 30.

191 *Spectator, Vol. VII, No. 512*: this issue of 17 October 1712 contains the story of a vizier, who, by pretending to understand the language of birds, reproached his war-mongering sultan by reporting that two owls had commented that they should never want for ruined villages; the sultan was so touched that he rebuilt what he had destroyed and lifted his tyranny over his people.

192 *historians assert*: Livy (59 BC–AD 17), in his *History of Rome*, described how a surprise attack on the Capitol by the Gauls was thwarted by the sacred geese of Juno alerting the Roman guards.

193 *chanticleer*: cockerel.

caitif: caitiff; a wicked, despicable wretch.

194 *outlets*: fields or enclosures attached to a house.

eye-trap: something to catch or deceive the eye.

heliotrope: a contrivance to mark the variation points of the sun's rising and setting; for the kinds of measurements White made with his heliotrope, see NJ, 26 January 1781.

195 *Baker's . . . mountains*: Richard Baker, *Chronicle of the Kings of England* (1643), 397, describes how Marcley Hill, thrown up by an earthquake in Herefordshire in 1571, began to 'travel', destroying all in its path; it 'walk'd' in this way for two days.

quaint: wise, knowing; skilled (especially with respect to the use of language).

196 *hundred*: a subdivision of a county or shire, having its own court.

  memorable . . . 1764: White noted in GK at the start of 1764, that there had
  been 'heavy rains, & vast floods' for most of the winter; on 31 January
  1764, he recorded, 'Prodigious inundations all over England, Holland &
  Germany.'

  woody hanger . . . leap: the dramatic landslide drew crowds, and on one day
  alone there were more than a thousand visitors from different parts of the
  country. White was in London at the time and on his return to Selborne
  on 25 March he was too ill to inspect the site immediately; his description
  of the event is based on that of his fifteen-year-old nephew Jack, who lived
  with White.

198 *furze*: gorse; a thorny evergreen shrub (*Ulex eurpoaeus*) that thrives on
  poor soils and waste land.

199 *taken . . . knowledge*: to be moved away from all that is familiar to an
  unknown place.

200 *Not many . . . turf*: White noted that he 'Put 10 field-crickets in the bank
  of the terrass' on 24 May 1765 (GK).

201 *dog-days*: the hottest time of the year; the period when the Dog Star rises
  and sets with the sun (usually reckoned 3 July to 11 August).

  When . . . writing: White noted the activities of the house-crickets that
  lived in his kitchen (see NJ, 22 April 1780, 9 June 1782, 4 June 1785), but
  there are no accounts of an infestation of the sort mentioned in the letter.

202 *visit*: White was visiting his aunt at Ringmer; his account of the uncover-
  ing of the mole cricket's nest appears in NJ, 6 June 1771.

204 *Brisson . . . l'echasse*: Mathurin-Jacques Brisson had published a six-
  volume *Ornithologia* (Paris, 1760); *l'échasse*: stilt.

  stuffed with pepper: there was much discussion in the eighteenth century
  on how best to preserve stuffed specimens, and most recipes included
  pepper (see *Phil. Trans.*, 60 (1770), 184–7, 311–13); White noted down the
  ingredients of a 'proper antiseptic substance for the preservation of birds'
  in NJ at the start of 1770.

  these birds . . . bird: White included a fold-out plate showing *Charadrius
  himantopus* in his *Natural History* (opposite p. 259); he noted in NJ, 28
  April 1779, that the long-legged plover is 'the most rare of all British birds'.

  The flamingo . . . birds: White had received a specimen of a flamingo from
  his brother John (that 'most accurate observer of Nature', mentioned later
  in the letter), which he presented to the Royal Society of London on
  26 March 1772.

205 *tortoise . . . property*: White took the tortoise, Timothy, to live at The Wakes
  following the death of his aunt Rebecca Snooke at Ringmer on 8 March
  1780; White and Timothy travelled to Selborne on 17 March.

206 *More Particulars . . . History*: in the first edition of *Selborne*, these particulars were published after the section on the 'Antiquities'.

Much . . . well: from Alexander Pope, *The Second Epistle of the Second Book of Horace Imitated* (1737), 13, line 191.

haha . . . precaution: in the 1750s White had followed the latest garden improvements and opened up vistas in his father's garden at The Wakes by constructing a haha, or deep ditch between the garden and the pasture beyond; a haha keeps livestock out of the garden while giving the impression that the garden extends into the surrounding countryside. White further expounded on the tortoise's wisdom in not falling into the haha in the letter he wrote in Timothy's name to Hecky Mulso, daughter of his close friend John Mulso (see *LL*, ii, 126–9).

'scald with safety': see William Shakespeare's *Henry IV, Part II*, Act 4, scene 5.

207 *miscellanies . . . deserve*: Barrington's *Miscellanies* (1781) covered a wide range of topics, from ornithology to Mozart; he acknowledged White in several essays, describing him as 'ingenious' and 'observant' (p. 225).

Fruit-walls . . . perpendicular: a work by Nicolas Fatio de Duillier, a member of the Royal Society; it was published in 1699.

208 *hippoboscae hirundinis*: flies that infest the nests of the swallow family (see the preamble to letter 16 to Barrington, pp. 128–9).

One swift . . . September: White's entry in NJ, 1 September 1782, reveals why this was remarkable: 'The swifts left Lyndon . . . about Aug^st 23. Some continued 'till Aug^st 29: & one 'till September 3rd!! In all our observations Mr Barker & I never saw or heard of a swift in Septem^r tho' we have remarked on them for more than 40 years.'

209 *Gibraltar . . . years*: John White had left Gibraltar in 1772.

nidus: a place in which the eggs of an insect can develop.

Lightfoot . . . Dorsetshire: in NJ, 29 May 1785, White noted that the vine was in the garden of Mr Le Coq of Weymouth; John Lightfoot had been vicar of Shalden, about six miles from Selborne, from 1765 to 1777.

210 *natural history . . . published*: in NJ, 29 May 1785, White referred to his brother John's 'excellent *Nat: History of Gibraltar*', which was in his possession following John White's death in 1780; Benjamin White, however, was dubious about the sales of this work, and it remained unpublished until 1911, by which point only the introductory part was extant.

Reaumur . . . for: see René-Antoine Ferchault de Réaumur's *Mémoires pour servir à l'histoire des insectes*, 6 vols. (1734–42), iv, plate 6, figs. 5–7.

attention . . . gardener: the gardener was Thomas Hoar; White noted in NJ, 31 May 1785, that 'Thomas persists in picking the *cocci* off the vine, & has destroyed hundreds', and on 6 June 1786, that the vine was once again infested with the *cocci*.

211 *family . . . bowl*: this was Henry White's family at Fyfield; goldfish were first imported into Europe from China in the seventeenth century, but did not become popular until the Dutch developed methods for their

artificial propagation in 1728. Most of the content of this letter can be found in NJ, 26 October 1782, during the time when White was visiting his brother.

211 *vivary*: vivarium; a glass bowl where fish are kept.

aliment: food, nourishment.

animalcula: plural of 'animalculum', the name given to the tiny creatures observed in drops of water when they were first examined under the microscope in the seventeenth century.

212 *corrupt*: to render impure, to taint.

stews: artificial ponds in which fish are kept until needed for the table.

215 *cotton . . . vegetables*: a down resembling cotton growing on some vegetable plants.

wimble: an instrument for boring holes, turned by a handle.

You . . . music: Barrington had employed his musical expertise to test the extraordinary skills of the eight-year-old Mozart and other musical child prodigies; see his *Miscellanies* (1781).

216 *Gassendus in Vitâ Peireskii*: Pierre Gassendi's biography of his friend Nicolas-Claude Fabri de Peiresc (1580–1637), a Provençal naturalist.

what . . . express: an allusion to Alexander Pope's poem 'An Essay on Criticism', *The Works of Alexander Pope, Esq* (1736), i, 117, line 300: 'What oft' was thought, but ne'er so well express'd'.

music . . . pleasure: White had made the same comment in a letter to Ralph Churton, following a large family gathering at The Wakes (from 6 September to 14 October 1782) when two of his nieces played the harpsichord 'day after day'; although White found the music 'very sweet and engaging at the time' it produced 'unpleasant sensations afterwards' (see *LL*, ii, 89).

nectarium: botanical term for the part of the flower that contains the nectar.

217 *house-swallows . . . upper-pond*: James Knight's ponds were located on the western edge of Selborne.

A rare . . . winter: the first two paragraphs of this letter were drawn from NJ, 11 May 1782.

218 *neighbour . . . Company*: Charles Etty (son of Andrew Etty, vicar of Selborne 1758–84) served as mate on several voyages for the East India Company; he had returned from Canton in May 1787. White's account of the Chinese dogs is drawn from his notes in NJ, 15 September 1787.

dogs . . . dumb: according to Antonio de Ulloa, *A Voyage to South America*, trans. John Adams, 2 vols. (1758), ii, 229, it was only the dogs of the island of Tierra (one of the Juan Fernández Islands), not all South American dogs, that did not bark.

219 *Ysbrandt Ides . . . prick-ears*: Evert Ysbrants Ides, *Three Years Travels from Moscow over-land to China* (1706), plate 4.

print . . . world: see plate 70, 'A Man of Kamtschatka, Travelling in Winter', in the atlas (vol. iv) of James Cook and James King, *A Voyage to the Pacific Ocean . . . in the Years 1776 . . . 1780* (1784).

220 *Stone-curlew . . . dark*: White mentioned the stone-curlews in NJ, 1 March 1788, the latest date that information from this source was incorporated into *Selborne*.

221 *little girl . . . him*: White refers to the words in Psalm 147: 9: 'and feedeth the young ravens that call upon him'; the little girl was White's niece Mary White (Molly), daughter of Thomas White.

Observationes de Aëre: published by John Huxham in three volumes in 1739, 1752, 1773. It appeared in translation as *Observations on the Air and Epidemic Diseases from the year . . . Made by Doctor Huxham, at Plymouth* (1759–67).

scuds: scud clouds are low ragged cloud fragments.

swivel guns: guns, or the smallest type of cannon, mounted on a swivelling stand so as to turn horizontally in any direction.

Hermitage: see note to p. 51.

222 *anathoth*: see note to p. 180.

gentleman . . . Newton Valence: Edmund White, White's nephew, had become vicar of Newton Valence, the adjoining parish to Selborne, in 1785; White also noted in NJ, 27 August 1786, that it was 'Mr White of Newton' who brought the swivel guns. This letter (and letter 59) was especially composed for *Selborne*, as White reveals in a letter to his niece Molly on 26 November 1787, in which he asks her to give it to her husband Benjamin White Jr (who was in business with his father Benjamin White) for insertion in the book before the letters to Barrington about the weather (see *LL*, ii, 171–3).

four following letters: at a late stage in the production of *Selborne*, White subdivided a long letter on winter weather into three but omitted to change this letter to read 'six following letters'.

223 *pease-haum*: the stalks of white peas, cut when green and dried quickly; used for fodder or straw.

224 *winds*: respiration; breathing.

froze . . . nights: as observed from the contents of chamber pots; the freezing point of urine is lower than that of water.

225 *declension*: gradual diminution or falling off.

226 *18th . . . over*: 18 January was Queen Charlotte's official birthday.

227 *rime*: hoar frost; frozen mist.

228 *A neighbour's . . . vigorous*: this hedge belonged to Richard Yalden, vicar of Newton Valence (see NJ, 22 February 1776).

230 *spiculae*: sharp-pointed crystals.

electric . . . people: displays of electrical phenomena were very popular in the eighteenth century, and usually included the transmission of a shock to a ring of people holding hands.

231 *old November*: in the old Julian Calendar, November fell eleven days earlier than it did after Britain adopted the Gregorian Calendar in 1752.

coddled: softened by heat.

honey-dews: the sugar-rich sticky liquid secreted by aphids.

232 *meteors*: atmospheric phenomena.

The summer . . . man: the effects were commented upon by other writers, such as William Cowper in his poem *The Task*, bk II (1785), 48, and Parson Woodforde in his diary (*The Diary of James Woodforde*, ed. R. L. Winstanley and P. Jameson, 17 vols. (1983), x, 146–63 *passim*); the cause of the phenomena described by White was the eruption of the Icelandic volcano, Skaptar Jökull.

234 *garden-lights and hand-glasses*: the glass placed over the frames on hotbeds, and the glass cloches used to protect tender plants.

235 *Mr Aikin . . . sort*: the third edition of John Aikin's *A Calendar of Nature* was published in 1787; the previous two editions had appeared in 1784 and 1785.

ACKNOWLEDGEMENTS

This edition has benefited enormously from the work of other Gilbert White editors and scholars, especially that of Paul Foster. I am particularly indebted to his earlier edition of *The Natural History of Selborne* in the World's Classics series.

I should like to thank Diana Donald, Ralph O'Connor, Miles Ogborn, Michael Hunter, and Shelley Innes for their comments and suggestions, and Rachel Iliffe, Isla Fay, and Beth Hannon for their help with establishing the text. Elizabeth Smith cast her expert eye over the Latin and Greek phrases and translations, and I am grateful for the revisions she provided.

It has been a great pleasure to work with Judith Luna, and I have much appreciated her enthusiasm, patience, and editorial judgement. Jim Secord has been an unfailing source of encouragement. My biggest debt is to him.

I wish to acknowledge permission from the copyright holders to include the following material in Appendix II:

W. H. Auden: 'Posthumous Letter to Gilbert White', *Thank You, Fog: Last Poems by W. H. Auden* (London: Faber and Faber; New York: Random House, 1974), 32–3. Copyright © 1974 by W. H. Auden. Reprinted by permission of Curtis Brown Ltd; 'Posthumous Letter to Gilbert White'; Copyright © 1973 by the Estate of W. H. Auden, from *Collected Poems of W. H. Auden* by W. H. Auden. Used by permission of Random House, Inc.

Francis Boott, letter to Charles Darwin, 22 December [1862], *The Correspondence of Charles Darwin*, ed. F. H. Burckhardt et al. (Cambridge, 1985–), x, 621. Reprinted by permission of the Syndics of Cambridge University Press.

Thomas Carlyle, letters to J. P. Eckermann, 2 August 1848 and 5 January 1849. Courtesy Duke University Press and *The Carlyle Letters Online*.

Samuel Taylor Coleridge, annotation in his copy of *The Natural History of Selborne*, *The Collected Works of Samuel Taylor Coleridge: Marginalia*, ed. H. J. Jackson and G. Whalley, 6 vols. (London: Routledge and Kegan Paul; Princeton: Princeton University Press, 1980–2001), vi, 149. Reprinted by permission of Taylor and Francis Books (UK) and Princeton University Press.

Charles Darwin, 'Recollections of my Mind and Character', *Charles Darwin: Evolutionary Writings*, ed. J. A. Secord, Oxford World's Classics (Oxford, 2008), p. 369. Reprinted by permission of Oxford University Press.

George Eliot, journal entry, 22–3 June 1855, *The Journals of George Eliot*, ed. M. Harris (Cambridge, 1998), 57. Reprinted by permission of the Syndics of Cambridge University Press.

Nathaniel Hawthorne, letter to E. A. Duyckinck, 1 July 1845, *Nathaniel Hawthorne: The Letters, 1843–1853*, ed. T. Woodson, L. N. Smith, and N. H. Pearson (Columbus, Ohio, 1985), p. 106. Reprinted by permission of Ohio State University Press.

D. B. Wyndham Lewis, 'Standing By . . .', *The Tatler* (20 August 1947), 242. Reprinted by permission of Mary Evans Picture Library.

H. J. Massingham, *The Writings of Gilbert White of Selborne* (London, 1938), pp. ix, xi–xii, xv, xxiv. By kind permission of Mrs Jill Spencer-Jones.

E. M. Nicholson (ed.), *The Natural History of Selborne* (London, 1929), 34–5. By kind permission of Piers Nicholson.

N. Nicholson, *Wednesday Early Closing* (London: Faber and Faber, 1975), 190. With grateful acknowledgement to the author and the publisher.

Henry Thoreau, response to a questionnaire, December 1853, *The Correspondence of Henry David Thoreau*, ed. W. Harding and C. Bode (New York, 1958), 309–10. Courtesy of New York University Press.

Raymond Williams, *The Country and the City* (London: Chatto and Windus, 1973), 118–19. Reprinted by permission of The Random House Group Ltd and Merryn Williams.

William Wordsworth, letter to F. Wrangham, 2 October 1808, *The Letters of William and Dorothy Wordsworth: The Middle Years, Part I: 1806–1811*, ed. E. De Selincourt, 2nd edn., rev. M. Moorman (Oxford, 1969), 270. Reprinted by permission of Huntington Library and Oxford University Press.

William Wordsworth, letter to H. C. Robinson, [? 24 May 1833], *The Letters of William and Dorothy Wordsworth: The Later Years, Part II: 1829–1834*, ed. E. De Selincourt, 2nd edn., rev. A. G. Hill (Oxford, 1979), 622–3. Reprinted by permission of Oxford University Press and the Trustees of Dr Williams's Library.

BIOGRAPHICAL INDEX

INDEX